Praise for the First Knowledges series …

‘This beautiful, important series is a gift and a tool. Use it well.’

—Tara June Winch

‘An in-depth understanding of Indigenous expertise and achievement.’

—Quentin Bryce

‘Australians are yearning for a different approach to land management. Let this series begin the discussion. Let us allow the discussion to develop and deepen.’

—Bruce Pascoe

‘These First Knowledges books are proving among the most fascinating and important titles I have ever read; an astounding gift of wisdom delivered with generosity and optimism, offering no less than a new vision of what Australia is, and what it can be … they deserve to change minds, lives, and hopefully the development of Australia itself.’

—Jez Ford

Danielle Gorogo, *Kinetic Synergies*, 2022

Kinetic Synergies is about the relationship between different types of energies in motion. Some energies when brought together achieve a synergy, allowing them to move towards a type of perfection that can only be achieved through collective effort and collaboration, to innovate and adapt, in an ever-changing world.

Danielle Gorogo is a Clarence Valley First Nations artist living in the Northern Rivers region of New South Wales. She is a direct descendant of the Dunghutti, Gumbaynggirr and Bundjalung nations. Danielle's multifaceted cultural heritage, which includes First Nations Australian, Papua New Guinean, Māori and Micronesian ancestry, is reflected in her art.

INNOVATION

Aboriginal and Torres Strait Islander peoples are advised that this book contains the names of people who have passed away.

The stories in this book are shared with the permission of the original storytellers.

INNOVATION

Knowledge and Ingenuity

IAN J McNIVEN &
LYNETTE RUSSELL

Thames &Hudson

First published in Australia in 2023
by Thames & Hudson Australia Pty Ltd
11 Central Boulevard, Portside Business Park
Port Melbourne, Victoria 3207
ABN: 72 004 751 964

thamesandhudson.com.au

26 25 24 23 5 4 3 2 1

Thames & Hudson Australia wishes to acknowledge that Aboriginal and Torres Strait Islander people are the first storytellers of this nation and the Traditional Custodians of the land on which we live and work. We acknowledge their continuing culture and pay respect to Elders past and present.

ISBN 978-1-760-76303-9 (paperback)
ISBN 978-1-760-76304-6 (ebook)

A catalogue record for this book is available from the National Library of Australia

This project has been assisted by the Australian Government through the Australia Council, its arts funding and advisory body.

Front cover: *Kinetic Synergies* by Danielle Gorogo

Series editor: Margo Neale
Cover design: Nada Backovic
Typesetting: Megan Ellis
Editing: Katie Purvis

Printed and bound in Australia by McPherson's Printing Group

FSC® is dedicated to the promotion of responsible forest management worldwide. This book is made of material from FSC®-certified forests and other controlled sources.

For Aunty Iris and Aunty Olga, storytellers, teachers.
They were our professors.

CONTENTS

FIRST KNOWLEDGES

MARGO NEALE, SERIES EDITOR

How could the First Peoples of an 8-million-square-kilometre continent of such topographic diversity, small in number, live here for 65,000 years and *not* be masters of innovation and invention? The Indigenous Australians ensured the vitality and resilience of their culture while keeping their environments healthy and sustainable.

The First Peoples had to adapt to a variety of climate changes. They survived the formation of new terrains brought on by successive ice ages, sea-level rises of up to 120 metres and the loss of vast tracts of land. The dramatic flooding of the continental shelf associated with global warming at the end of the Pleistocene (around 12,000 years ago) saw around one-quarter of the continent inundated. Inland cultural groups had to adapt to becoming coastal people, and vice versa. These were huge adjustments given the culture is embodied in the environment.

But innovations were not just fashioned for tangible events. In this book, authors Ian J McNiven and Lynette Russell show that First Nations communities accommodated changing social as well as environmental circumstances. These included the adaptations communities made in response to the arrival of waves of outsiders, in particular those who imposed their cultural dominance on our ancestors. The authors disabuse us of the naive notion that throughout tens of millennia Aboriginal nations stayed unchanged. They challenge such opportunistic narratives promoted by successive generations of colonisers, which were contrived to ensure the ownership of lands never ceded by our First Peoples.

Ian and Lynette are long-time collaborators who have worked with communities across the continent for decades. Together they bring to this topic a rare depth of knowledge and experience, working from both historical and archaeological perspectives. They show that Indigenous innovation is not linear, encompassing only the new, but must be seen as a holistic approach that draws widely upon existing and inherited social and cultural beliefs and practices. They counter the old view that First Nations people were 'conservative and tradition-bound ... passive rather than progressive', immune from change.

Rather, they demonstrate First Peoples' capacity to accommodate and incorporate relevant elements from other cultures – from outsiders and from other Indigenous and Torres Strait Islander groups. We live modern lives. Today we have innovatively assimilated all aspects of Western knowledge and participate widely in, for example, the arts, sciences, fashion, technology, agriculture and aquaculture. Innovation and inventiveness have been at the core of First Nations cultures for at least 65,000 years.

Ian and Lynette dispense with the use of the term 'traditional'. For them, it counters their experience of a dynamic and continuous culture, casting, as they say, 'a negative shadow over contemporary Indigenous culture, allowing it to be readily perceived as somehow diminished, or even inauthentic'. Their work with Indigenous communities over the past thirty years highlights the way new raw materials, ideas and concepts have, as Ian and Lynette describe it, been seamlessly incorporated into Indigenous cultural repertoires. New trading systems, new systems of clan organisation and new intergroup ceremonial gatherings are a few demonstrations of this. Today, most Australian archaeologists acknowledge that cultural change 'is a result of a complex mix of environmental and social changes, each associated with innovation and accommodation'.

By sifting through the mounds of a history extending back millennia, these authors reveal some compelling findings. Excavations at Madjedbebe in Arnhem Land revealed the use of ground-edge axes dating back between 55,000 and 65,000 years. Though also found in other Ice Age sites across the north, these ancient Arnhem Land axes are the oldest known examples in the world of people hafting handles onto tools. Further, we discover the sophisticated watercraft skills of the ancestors of modern Aboriginal peoples, who were the 'world's first mariners'. They sailed to Australia from South-East Asia at least 65,000 years ago, navigating 100 kilometres of sea in open vessels.

The *Innovation* authors tell us about the Old People of Ice Age Australia who developed some of the world's earliest known expressions of humanity: the adoption of practices relating to spirituality and symbolism. Sophisticated rituals around ancient burials date back tens of thousands of years. For example, the Lake Mungo human burials in New South Wales are dated to about 40,000 years ago. Here a range of mortuary practices, including cremation and full-body burials, have been identified by anthropologists. The 40,000-year-old cremation of Mungo Lady and the ochre full-body burial of Mungo Man, together with the 20,000-year-old cemetery at Kow Swamp in central-northern Victoria, are considered among the earliest known examples of mortuary practices. These are watershed examples for the reassessment of the history of Aboriginal Australia and the ingenuity of our First Peoples.

The complex trading networks of the continent as described by Ian and Lynette were geographically extensive, highly effective, and in many cases can be likened to religious pilgrimages. Australian Indigenous innovation was in play and evident on all counts, not only through the geographical scale and volume of materials traded, but also through the ways in which the social and cosmological dimensions of our culture were applied. Ceremonial objects traded involved Ancestral Law and were used for strengthening relationships between communities. The celebration of religiously ordained (Ancestral Law) relationships between communities and these trade networks was enhanced by changing social alliances. Indeed, the boundaries between gift-giving and trading were often indecipherable to the outsider.

Torres Strait Islanders also developed an elaborate trading system, based on the movement of canoes as the vehicles that carried items of trade. Routes extended north to include New Guinean Melanesians and south to engage with Cape York Peninsula Aboriginal peoples. Trade objects and ideas were exchanged over thousands of kilometres and often traversed hundreds of clan territories and one side of the continent to another.

The seasonal calendars created thousands of years ago and used today not only secured resources but also signalled the timing for ritual and cultural activities. In this sense, as the authors observe, the staging of cultural events was tuned to an understanding of seasons and their impact on the environment.

The authors convincingly replace a Western view of Aboriginal people as hunter-gatherers living at the mercy of the environment with one of agency and an understanding of the interdependencies of all living things. This is built on a philosophy that says if you look after the environment, then the environment will look after you. Such a mutual relationship and co-dependency, energised by Ancestral powers and spiritual life forces as the authors explain, is at odds with the Western view of Country as being inanimate and filled with resources for exploitation. A well-known example of collaboration with nature is 'firestick farming', a controlled cultural-burning practice that cleared tinder-dry old growth from forests during the winter and encouraged the new in spring. It transformed landscapes across the continent, allowing plants and animals to thrive. This capacity came from deep-time ecological knowledge gained through thousands of years of intimate observation and empirically based scientific understandings, and being instructed by ritual and ceremony.

Aboriginal First Peoples devised food-preparation processes for making otherwise toxic plants edible. Their innovation in aquaculture in freshwater eel and saltwater oyster farming is some 6600 years old. Fish traps were engineered all around the continent. They included intricately constructed networks of stone-walled channels and ponds to assist the navigation and farming of eels across larval outcrops, and fish traps of various construction in inland rivers and marine environments. The peoples' ecologically engineered practices are being used today to enhance sustainable oyster farming and other marine life, augmenting Western scientific approaches to land and sea management.

Our First Peoples developed relationships with marine animals – dugong, turtles, whales, dolphins and suckerfish – to assist in hunting and fishing. Suckerfish are tied to a string line and, once released from a boat, can attach themselves to turtles, fish or dugong to aid their capture by fishers. Hunters 'sing' (ritually call up) dolphins to assist with fishing. Historically, the killer whales or orcas near Eden on the south coast of New South Wales worked cooperatively with whalers to herd other species of whales to the shore for the whalers to kill.

In the pastoral settings of colonisation, displaced Aboriginal people dealt with the newcomers in very proactive social and cultural ways. Some worked alongside them to ensure they would retain connection to Country, becoming stockmen/women, trackers and native police. They adjusted to the shifting social and economic circumstances while retaining agency, enterprise and entrepreneurship. This was not, however, without drawing on their own capacities to effectively defend themselves. In the whaling industry, Aboriginal men used their outstanding harpooning skills with spears to gain some economic independency and agency.

The authors show us that what the Europeans termed 'begging' was one of the truly innovative economic strategies used by Aboriginal people to engage with the imposed economic system. 'Begging' was a practical means for obtaining money and other commodities, an innovative and justifiable tactic that expressed the First Peoples' right to material goods and money in compensation for what the Europeans had taken from them. It was reciprocal trade for their eviction and relocation, and disruption of their customary hunting practices. Given it lasted for decades, 'begging' was clearly an effective strategy for securing what was wanted and was consistent with the customary practices of reciprocity and gift-giving.

The rock-art imagery of northern Australia reveals a remarkable example of Indigenous sovereignty. The proliferation of images of European ships (and horses and guns) are figured to indicate the European presence in the region as an imposition. For centuries before Europeans arrived in Australia, Makassan mariners visited the northern coastline and gathered timber, shells from turtles, and pearls while fishing for the highly lucrative trepang. They were seen as allies and trading partners. By contrast, as the authors explain, rock-art imagery implies that Europeans were viewed as invaders, usurpers and colonists. There are many fewer depictions of Makassan ships, potentially indicating friendly, negotiated cultural exchanges over a very long period.

The cross-cultural interactions with the seasonally visiting Makassans reinforces the innovative mindset of the local peoples. They adopted Makassan technologies, including the dugout canoe and various dwelling designs. They stocked Makassans' bamboo pipes, tobacco, cloth, belts, string, alcohol, rice and tamarind. They also obtained ceramics, glass and metal objects, which they valued highly, adapted and often carefully maintained. However, Aboriginal people were highly selective about what they incorporated into their cultures. For ceremonial purposes, the shape of the furled and unfurled sails of the Makassan vessels called *praus* were adopted as totems, and songs were sung indicating the north-west and south-east winds that brought the visitors annually to and from the northern shore. These are still sung today.

The adoption of new materials and techniques into artistic practices did not result in the rejection of previous art practices but, rather, enhanced what people were already doing. Found glass from the Europeans was flaked into stunning spearheads, now considered to be the pinnacle of Aboriginal ingenuity and material culture. In the 19th century Reckitt's Blue, a household laundry product, began to be incorporated by Aboriginal painters to enhance their colour palette for rock art and many other artefacts. The use of blue continues today on bark paintings in Arnhem Land, though with paint, not Reckitt's. In the past decade, the proliferation of ghost-net art has seen the use of discarded fishing nets to make sculptures of great artistic merit that are popular around the world. The nylon and polypropylene fibres are knotted and woven using traditional techniques. Also textile-based, they are innovative applications of Indigenous motifs in fabric design to reinforce identity, heritage and political sloganeering on all manner of garments. Here we see a great demonstration of Country and connectivity: just as artists see themselves as instruments for channelling the ancestors, so do the innovators.

Innovation: Knowledge and Ingenuity is an illuminating addition to the First Knowledges books. The first in the series, *Songlines: The Power and Promise*, establishes the foundational truths about how all knowledge resides in Country, including astronomy, medicine, engineering, ecology, kinship systems and social mores. *Design: Building on Country*, the second book, explains the importance of building as an extension of Country and designing spaces as a collaborator, not a usurper. It shows how we invest objects made from Country with the spirit of our ancestors. The third book, *Country: Future Fire, Future Farming*, is a timely call to action for a conversation about who we are as Australians on this continent that has been so rapidly exploited in recent generations, and about how to take responsibility for its restoration. The fourth book, *Astronomy: Sky Country*, directs our gaze upwards and reminds us of the importance of referencing inherited cosmologies and ways of being. As with all things in an integrated knowledge system, *Sky Country* restores the connectivity between sky and earth that modern peoples have lost sight of. The book also invites a timely conversation about how the degradation of our environment has an equally corrosive impact on our Sky knowledge system. The authors proffer some achievable solutions that can guide us towards taking responsibility for the restoration of dark skies. The fifth book in the series, *Plants: Past, Present and Future*, celebrates the deep cultural significance of plants and shows how engaging with this heritage could be the key to a healthier, more sustainable future. *Law: The Way of the Ancestors*, the sixth book, offers insight into how Indigenous law can inspire new ways forward for us all in the face of significant global issues. It states that Indigenous law defines what it is to be human, and offers a path to resilient and caring communities.

This, the seventh book, directs our gaze more downwards and outwards, because everything is connected. Like other books in the series, it provokes bigger national discourses on the innovativeness of First Peoples – and indeed all Indigenous peoples of the world – and about the critical relevance of Indigenous wisdoms right now. First Peoples' inventive capacity is there in the resourceful and dynamic ways we have faced enormous challenges to our worlds, and, further, in the ways we have accommodated differences and survived under the most oppressive circumstances. It is in the lessons from these histories of inspirational self-determination that the contemporary world will find answers to the imminent and catastrophic threats that face it.

A feature of this series is that each volume concludes with a vision of the future. Here, Lynette and Ian advocate a positive turn for archaeological practice – namely, recognising that Indigenous peoples were always innovative and adaptive and that our culture is dynamic and sophisticated – and for an Indigenous Voice and self-determination in matters concerning Indigenous peoples. In so doing, they acknowledge the deliberate erasure of Indigenous agency and voice in past histories and narratives as a strategy in support of the colonial project to dehumanise Indigenous peoples of Australia. They have repurposed the slogan coined by disability activists 'Nothing about us without us' and claimed it as a directive for us to be heard and observed in this field. Custodians of the land, as Indigenous peoples are, must be consulted and must lead in archaeological and historical projects.

Innovations discussed in this book relate to social, religious, economic, political, ecological and cultural contexts. They highlight sophisticated complexities in the culture that have often gone under the radar. In all cases, such innovations were underpinned by ancient, spiritually inspired philosophies that placed primary value in social relationships between people, and between people and the land. Emphasising sustainability, such actions go to the heart of the concept of Country and its embodied knowledge.

In the challenging environmental circumstances of this fragmenting planet, it appears that the oneness with nature that Indigenous people have is much sought after. If the peoples of the Western world wish to survive, it is critical that they start looking to the ancient wisdoms of the First Peoples of the world, where they will find a rich resource to assist in the process of planetary healing.

The present is catching up with the past.

1

PERSONAL PERSPECTIVES

IAN J McNIVEN

I was one of those rare kids who always knew what they wanted to be when they grew up. Ever since I can remember, I wanted to be an archaeologist. Growing up in a middle-class suburb of northern Brisbane seems like an unlikely place to nurture archaeological aspirations. Yet my parents, Peter and Glenda McNiven, never doubted my desire and always encouraged me. I'm not sure they thought being an archaeologist in Australia was viable, but they always supported my passion and drive. I remember Dad telling me that his father, Ronald McNiven, the well-known ice-cream and confectionary manufacturer in Sydney, thought it best I get a real job and practise archaeology as an intellectual hobby. Ever optimistic, I took that advice as 'Make my real job archaeology'.

As a young kid in the late 1960s and early 70s, a great source of knowledge on archaeology and 'prehistoric man' was How and Why Wonder Books.[1] In 1977, when I was a teenager, my paternal grandfather died and I inherited his copy of *The Epic of Man*, published in 1961. This coffee-table book by the editors of *Life* magazine was filled with the most extraordinary paintings of reconstructions of life in the prehistoric past, complete with colourful scenes of Neanderthals hunting a mammoth and an elaborate Upper Palaeolithic burial ceremony in a cave in France.[2] Here was the prehistoric past brought alive in all its ethnographic detail. The final section of the book surveyed contemporary peoples for 'analogies' of the past who were 'left behind, in stagnant cultural pools'. Aboriginal Australians were featured across six colourful pages as an example of 'Stone Age Cultures Today'.[3] Despite its dated language and concepts I still treasure this book, and looking at it as I write this paragraph brings back childhood memories and dreams. It also makes me think that perhaps this is where I learnt that the past is as much about anthropology as it is about archaeology, because it is always about the people behind the artefacts.

Fast forward to 1980 and my first year at the University of Queensland as a fresh-faced kid straight out of school. In my last year at high school, Dad contacted the uni to see what options existed for me to become an archaeologist. Luckily, he was put through to the Anthropology department and not Classics and Ancient History. It seemed clear that the best route to get employed as an archaeologist in Australia was to study archaeology as a subdiscipline of anthropology. That meant 'prehistoric archaeology', which was a big thumbs-up from me, and Australia's Aboriginal past, which left me wondering a little. After the first few lectures I was hooked to the point of becoming somewhat of a zealot – my poor parents! Jay Hall was my lecturer in Archaeology and it is thanks to his teaching and mentorship that I eventually received a PhD in Australian Indigenous archaeology (technically my PhD certificate says 'Doctor of Philosophy in the Field of Anthropology', which I wear as a badge of honour).

Looking back now, I can't believe that the first friendships I made with Aboriginal people were at the University of Queensland. Here were people who, like me, had lived in Brisbane all their lives, but my knowledge of First Nations Brisbane was next to zero before my uni days. An important lesson I learnt at uni was to never let your studies get in the way of your education. It was here that I began to understand that to be a good archaeologist working with First Nations communities, one requires a good trowel and a good set of ears. I learnt lots of new information from my textbooks, but I also learnt lots by chatting to Aboriginal people. Many of these conversations took place in the tearoom of the uni's Anthropology Museum. In the 1980s the museum, directed by Peter Lauer, was at the cutting edge of involving First Nations community members in anthropological, historical and archaeological research. The most important lesson I learnt was that no matter where I undertook archaeological research in Australia, I would be working on Country and the heritage of a particular Indigenous community.

It is their heritage and their story to tell, and if I am lucky then my archaeological skills may be useful in this regard.

After graduating with my PhD, I was employed as a postdoc researcher in archaeology at La Trobe University in Melbourne. The best thing to come out of that short experience was that I met Lynette Russell, who eventually became my soulmate, wife and research collaborator.[4] We are the greatest supporters of each other's research, but equally important is that we also provide honest critiques. Our key research overlap is understanding the colonial foundations of Australian archaeology and its contemporary legacies in academic and public arenas. We soon found that Australia's colonialist archaeology had much in common with the USA and Africa: it was a weapon of dispossession in settler colonies. Early on, we detailed this research in our 2005 book, *Appropriated Pasts: Indigenous Peoples and the Colonial Culture of Archaeology*. Regrettably, such colonial views continue to rear their head from time to time; a recent example is the debate over the origin of the Gwion Gwion rock paintings (the so-called 'Bradshaws') in the Kimberley region.[5]

In late 1992 I was offered a job as Cultural Heritage Officer with Queensland National Parks for the newly declared Fraser Island World Heritage Area, and I moved to Maryborough. An important part of that job was working with the more than ten Indigenous organisations that had interests in the region. Boy, did I have a steep learning curve in community politics and the legacies of colonialism in the management of cultural heritage sites and places. It was during my one-year stay in Maryborough that my friendship with Butchulla Elder Aunty Olga Miller strengthened. Aunty Olga became my new teacher and mentor as we chatted for hours at her kitchen table drinking endless cups of tea. One of my treasured possessions is a signed copy of Aunty's *Fraser Island Legends*, published in 1993 when she was aged seventy-three. Inside the front cover she wrote: 'Best wishes Ian, for the future & don't forget to keep "digging"'.

In 1996, my archaeological education took another leap forward. I was invited by anthropologists John Cordell and Judith Fitzpatrick to participate in a cultural heritage overview of Torres Strait in Far North Queensland. I accepted the invitation knowing very little about Torres Strait Islanders. Judith and John introduced me to the Goemulgal community of the island of Mabuyag (aka Mabuiag) in central-western Torres Strait. I met senior Elders and became friends with a number of community members, such as Cygnet Repu, Terrence Whap and Sophie Luffman, who were the next generation in line to assume leadership roles. Here, my archaeological skills on shell middens and reconstruction of ancient diets were not as impressive as I thought. It soon became clear to me that when it came to their history, the Goemulgal were less interested in marine subsistence specialisation (they already knew that as they lived it every day!) and more interested in their social and ceremonial history.

After giving myself a crash course in 19th-century Torres Strait Islander culture by reading the six volumes of Alfred Haddon's *Reports of the Cambridge Anthropological Expedition to Torres Straits* (1901–35), it dawned upon me that a key dimension of the specialised maritime culture of Torres Strait Islanders is totemic and ritual relationships with the sea. I realised that my theoretical toolkit needed some serious innovation if I was going to undertake archaeological research that was relevant to the Goemulgal. As a result, I developed the notion of 'seascapes as spiritscapes' and, together, the Goemulgal community and I fruitfully researched the archaeology of totemic ceremonial sites (known as kod sites) and dugong-hunting rituals.[6] My skill set in seascape archaeology continued to sharpen through a collaborative research partnership with the Kulkalgal of central Torres Strait, focusing on the history and archaeology of the reef island of Tudu. All of these newly acquired skills helped me complete my 2008 Expert Witness report on archaeology on behalf of the Torres Strait Native Title Office for the Torres Strait Sea Claim. Here, I had an opportunity to give back to all Torres Strait Islanders and to put into action the community benefits and relevance of Indigenous archaeology.

As a professional archaeologist for nearly forty years, I have worked with more than fifty Indigenous cultural groups in Australia and Papua New Guinea. I have never stopped listening to Elders to learn new skills to better my qualifications to work with and for Indigenous communities and to successfully co-design collaborative archaeological projects. Not all First Nations communities are interested in archaeological research but most are, as long as it is done the 'proper way'. It is not hard for me to keep my interest in undertaking archaeological research with First Nations communities strong. Apart from making new friends and always having a good laugh, I find the Indigenous history of Australia truly remarkable and fascinating. It is a great honour to be in a position where I can work with and for First Nations communities to research their history and heritage through archaeology. Yep, my real job is archaeology.

LYNETTE RUSSELL

My journey, like most, needs a bit of context. Growing up in Melbourne's outer suburbs in the 1960s and 70s meant being part of a large multicultural community, mostly made up of first-generation Italian, Greek, Cypriot and Turkish families. My high school had over 1500 students, most from non-English-speaking households. Education was to be endured, and leaving school – usually at around 15–16 years – was celebrated. For the lucky ones, a hairdressing or fitting-and-turning apprenticeship represented freedom. When the local Westfield shopping centre opened, many of my classmates opted for jobs in retail at Kmart or Target. Most people's aspirations were modest.

At school I loved history and literature, two things my teachers assured me would never lead to a job – other than, possibly, as a teacher. Our misnamed 'careers adviser' only ever spoke of 'jobs' – 'careers' were not for us. No one in my family had finished high school, let alone studied at university. We didn't talk much about history, heritage or culture. Like many parents, mine were focused on paying off the mortgage and keeping my brothers and me fed and clothed. My father, grandfather and, for a time, great-grandfather would, on most weekends, go fishing, or ferreting for rabbits. They would often bring home mushrooms with their haul. Very occasionally I was allowed to accompany them. It was on one of these trips that I first tasted wattle sap, taken directly from a black wattle tree. The sap, just a tiny piece placed on my tongue, was lemony sweet. At the time I had no idea that I was being shown knowledge that had been handed down the generations from my Aboriginal grandmother's mother's mother, shared with partners, husbands and children.

My understanding of Aboriginal culture was gained slowly at first, but for the past forty years I have dedicated myself to learning everything I can. Thirsty for knowledge, I studied history, archaeology, anthropology and ethnography and completed a PhD at the University of Melbourne in the mid-1990s. For my PhD I looked at the ways in which popular culture and academic writings depicted Aboriginal people up to the European Bicentenary in 1988. I became aware that for the most part Aboriginal people and culture had been homogenised and that tribal, group and geographic differences were largely ignored or forgotten. My first academic work tried to emphasise that we need to seek out regional and cultural specificities to show the amazing diversity of a continent with over 250 different languages and more than 600 different 'tribes'.[7]

Keen to understand the diversity of Indigenous Australia, I worked in various Aboriginal community heritage and educational organisations. In 1996 Ian J McNiven and I both began work with the Victorian Native Title Unit. Previously I had worked at Aboriginal Affairs Victoria in the Archaeological Survey unit (VAS), where I became immersed in the Victorian Koori community and formed friendships and relationships that have lasted ever since. Working at VAS was life changing – I was able to explore my family's heritage and look at connections and disconnections across the state. Luckily for me, VAS had an extraordinary library of virtually everything that had ever been written on Victorian Aboriginal people and cultures; it was a treasure trove for the curious.

However, it has been less the book work and more 'fieldwork' or, perhaps more correctly, Country work that has truly educated me. One of the highlights of my work at VAS was being able to go out into regional Victoria to explore the land and see the heritage up close. In this context I was able to walk around sites with Traditional Owners and see firsthand how they engaged with and belonged to Country. It was on these field trips out in the western Victorian arid zone that esteemed Wotjabaluk Elder Uncle Jack Kennedy took the time to teach me how to listen, look and feel. He and his brother Uncle Patrick had known my great-grandmother and showed me places that had been important in her life, places our family had forgotten over time. Uncle Patrick had lived almost all his life on Country. His neat-as-a-pin corrugated-iron home had a soft sand floor – sand that had come from the Dimboola River. He had no electricity, no refrigeration, his battery-operated radio kept him informed and Tilley lanterns were his only source of artificial light. Uncle Patrick cooked on an open fire and lived a life as close as possible to that of the Old People. While others might have seen poverty, I saw dignity. He lived this way until shortly before he died, on the Country he loved and where he belonged.

A little to the south, Kerrup Jmara Elder Aunty Iris Lovett-Gardner encouraged me to write it all down, to find connections and to make relationships. Aunty Iris had been instrumental in setting up the Koorie History Unit in Fitzroy and was one of the first people I met when I started my journey. She had spent time with many people from the Stolen Generations and understood what it felt like to be searching for answers.[8] Aunty Iris told me about Bill Onus's

Aboriginal Enterprises in Belgrave, a tourist venture where busloads of visitors would arrive and Bill would make boomerangs and other traditional wooden artefacts. Aunty Iris had been his salesperson, a job she loved to talk about even years later. I was inspired by her dedication and her never-ending commitment to making things better for the community. Perhaps most of all, I was honoured by her welcoming of me and my search.

While working at Mirimbiak Native Title Services, Ian and I encountered the larger-than-life figure of Richard Frankland, Gunditjmara wise man, activist, filmmaker, songman. Richard taught me to ask questions, to seek answers and then allow them to wash over me. He did Ian and me the honour of being best man at our wedding in 1999 and offered an Acknowledgement of Country, which was a first for many in the room. Uncle Jim Berg and Aunty Joy Murphy-Wandin, and other Elders from Naarm (Melbourne), where I have always lived, have encouraged me to continue to explore, and to share what I learn. Over the decades myriad students – undergraduate and postgraduate – have permitted me the chance to 'pay it forward', giving back and making a difference. Over the past three-and-a-half decades I have been privileged to have had amazing teachers.

When contemplating the concept of 'innovation' as it relates to Indigenous Australians, two words echo in my mind: resilient and resourceful. Resilient goes without saying, when we see survival in the face of incredible odds. Resourceful is something I have seen repeatedly in various situations. I recall talking to community members about what it was like in the 1950s and 60s when the 'welfare inspectors' would come and visit. Everyone was expected to have vegemite, cornflakes and milk in their house – it was required by the authorities. One old aunty said to me, 'Those inspectors were going out the front door and the groceries were going over the back fence to the next house.' Innovation is at the very core of First Nations being and living.

Historians generally rely on written evidence; occasionally oral testimony might be used, usually for understanding more recent events. As a historian I use whatever clues I can find to tell the stories of the past. I have looked at texts, songs, museum collections, art and archaeological finds. Each strand of evidence gives me a little bit more that I can weave into a nuanced picture of the past. This kind of history writing has led to many fruitful collaborations, particularly with archaeologists. Writing this book with archaeologist Ian J McNiven is both a professional and a personal pleasure. As my life partner, Ian has played a key role in all my pursuits over the past thirty years, and as colleagues we share a deep desire to tell the stories of First Nations Australia. Our driving passion has always been social justice and a determination to highlight the complexity of the Indigenous past as well as the way that past affects the present. All of our collective efforts have been devoted to revealing the underlying colonialism and the degree to which our chosen disciplines of history and archaeology have been complicit. This has from time to time put us at odds with more conservative colleagues.

Ian and I wrote this book at our home in Melbourne, 'Lillimur', which means wattle sap in the Wergaia language and is also the name of the small town in western Victoria where my Aboriginal great-grandmother was born. Her resilience and resourcefulness remain a guiding light in my life.

OUR TAKE ON INNOVATIONS

When Europeans first arrived in Australia, there were hundreds of different cultural groups and over 200 language groups. European observers tended to conceptually erase this diversity of Indigenous cultures, leading to depictions of an 'essentialised, spatially homogenous Aboriginal culture'.[9] It is all too simple to perceive this spatial homogeneity as chronological stagnation, but together with significant environmental and climatic shifts, these cultures also saw significant cultural modifications. Volcanoes have erupted, dune fields have developed, glaciers have melted, and sea levels have risen by around 120 metres since people have lived in Australia. It is naive to assume that throughout tens of millennia Aboriginal nations have stayed unchanged.[10]

In this book we use the term 'Indigenous innovation' to refer to interesting, ingenious and original ways in which Australia's Indigenous peoples have responded to specific challenges, be they environmental, cultural, social, or the result of internal or external influences. One of the terms we avoid using as much as possible is 'traditional' as it tends to cast a negative shadow over contemporary Indigenous culture, allowing it to be readily perceived as somehow diminished, or even inauthentic.[11] As both of us have worked with many different Indigenous communities over the past thirty years, we have frequently been fascinated to see the ways in which new raw materials, ideas and concepts have been seamlessly incorporated into Indigenous cultural repertoires.

Our understanding of what constitutes an innovation is infused with our long-term relationships with Indigenous people from across the Australian mainland and islands. Yet during the late 19th century and the early part of the 20th century, the Western academy was less generous in its assessment of Australian Aboriginal innovation. Early historians and archaeologists tended to depict Indigenous Australian cultures as inherently conservative and tradition-bound. Culture, particularly Aboriginal culture, was often thought of as passive rather than progressive. Indeed, innovation was thought to be the antithesis of Australian Indigenous cultures. These cultures were seen as immune from development. Such representations were used as post-hoc arguments to explain why Aboriginal peoples were time-locked to an ancient past and, indeed, were living fossils.[12] In this sense, the 'living fossils' idea was a conclusion in search of an explanation. This colonialist and racist representation of First Nations Australians gave rise to the anthropological paradigm of diffusionism, whereby all complex and sophisticated cultural traits of Aboriginal Australians were explained away as overseas imports.

In his 1940 article on the origins of Aboriginal material culture, ethnologist and archaeologist Fred McCarthy at the Australian Museum in Sydney turned the diffusionist paradigm into an art form.[13] So extreme were McCarthy's views that one of his research collaborators, Herbert Noone, was moved to write: 'So much evidence has been put forward to indicate that the Australian aboriginal has brought, or borrowed, many traits of his culture from overseas sources, that one is led to look for anything of his that is left.'[14] In marked contrast, Noone spoke of the 'inventive genius' of Aboriginal Australians.[15] As late as 1977, McCarthy held the view that 'There is not a great deal of evidence to support invention as a prominent principle of Aboriginal culture'.[16]

Hyperdiffusionist and anatomist Grafton Elliot Smith, who believed the Egyptians were responsible for advanced cultural traits across the ancient world, was even less subtle in his pronouncements about Aboriginal innovations. In 1930 he wrote: 'Of constructive efforts of originality there is little or no trace. Apart from implements of stone, bone, and wood, they invented nothing.'[17] Anthropological and archaeological research over the past sixty years, examples of which are cited in this book, reveal the empirical fallacy of the diffusionist views of McCarthy and Smith. Such views can be consigned to the dustbin of history.

Another insidious expression of the conservative view of Aboriginal cultures was that the cultures were historically static and didn't change through time. If these cultures didn't change, then Ian would be out of a job as an archaeologist! Researching and understanding cultural change is inherent to the practice of archaeology. What people did 1000 years ago is rarely the same as what they did 5000 or 20,000 years ago. However, this is exactly what Western researchers thought about the Aboriginal past in Australia during the late 19th and early 20th centuries. Because Aboriginal people were thought to be an unchanging people in an unchanging land, their cultures were seen as static and locked in a prehistoric time warp. This timeless past left no place for archaeology, and indeed is a key reason why Australian Indigenous archaeology had such a late start. Why dig up archaeological sites when the same evidence is more easily available, and more complete, by making ethnographic observations today?

Theoretical cracks did emerge from time to time. For example, early ethnographic research into Victorian Aboriginal oven mounds included excavations where it was noticed that none of the sites appeared old. One excavator, the Reverend Peter MacPherson, wrote in 1885 that such evidence pointed to 'the building of ovens and the accumulation of the mounds of ashes as comparatively modern innovations'.[18] Unfortunately, these views on cultural change and innovation fell on deaf ears and had to wait until the 20th century to blossom. The big breakthrough occurred in 1929, when Norman Tindale and Herbert Hale of the South Australian Museum undertook the controlled excavation of deeply stratified deposits in Ngaut Ngaut (Devon Downs) rock-shelter on the lower Murray River in South Australia. Their excavations revealed a series of differing cultural layers, each containing different types of stone tools. Although Tindale and Hale didn't know the age of the layers (radiocarbon dating wasn't invented until the 1950s), they demonstrated that cultural practices of ethnographic times differed from those in the ancient past.[19] Finally, Indigenous archaeology had come of age in Australia and the doors were opened to find cultural change in other parts of the continent.

The idea that Australian Indigenous communities changed over a timescale measured in thousands of years only partly resolved the issue of seeing the Aboriginal past as static. One of the consequences of representing Indigenous peoples as living close to nature and even 'in harmony with nature' is that historical change in the past was tied closely to environmental change. That is, Aboriginal culture remained stable until faced with environmental change. When environmental change occurred, Aboriginal societies were seen to respond to such change (termed 'adaptation' in the 1960s and 70s) in an almost mechanistic fashion. As historian Billy Griffiths and Lynette Russell observe, the notion that Aboriginal peoples lived 'in relative harmony with each other and their environment' extended to perceptions that they 'achieved a society that had no need to change until Europeans arrived'.[20]

In the 1980s, archaeologist Harry Lourandos shook the foundations of Australian archaeology by questioning its overemphasis on linking cultural change to environmental change.[21] Lourandos pointed out that not only were such environmentally deterministic views anthropologically simplistic, but numerous examples existed of archaeological sites in Australia that documented cultural changes at periods in the past when little or no environmental change had occurred. Alternatively, Lourandos introduced more socially oriented theories and argued that many cultural changes in the ancient past were a result of internally generated social changes and innovations, such as new trading systems, new systems of clan organisation and new intergroup ceremonial gatherings. His approach was fiercely debated by Australian archaeologists, with many detractors holding tightly to cherished environmental theories of cultural change.[22] Today, most Australian archaeologists acknowledge that cultural change is a result of a complex mix of environmental and social changes, each associated with innovation and accommodation.[23]

First Nations people of Australia (Aboriginal peoples and Torres Strait Islanders) have thrived across their lands and seas for thousands of years. Over the course of at least 65,000 years, the number of Indigenous peoples of Australia went from a few hundred to at least 1 million at the time of British invasion in the late 18th century.[24] Over those 65,000 years, First Nations Australians engineered a unique set of cultures and societies for a diversity of environments, ranging from the sandy deserts of Central Australia to the snow-covered mountain ranges of south-eastern Australia and the warm tropical archipelagos of the Great Barrier Reef and Torres Strait. Such environmental diversity came with major challenges during periods of major climatic and environmental change, exemplified well by the peak of the last Ice Age around 20,000 years ago. Perhaps more dramatic was the flooding of the continental shelf associated with global warming at the end of the Ice Age, when around a quarter of the continent was inundated.

Across this vast expanse of time and dynamic environmental change, First Nations communities from different parts of the continent went through 'phases of innovation' to accommodate their own changing social and environmental circumstances.[25] Innovations ranged from changes in social and religious practices and institutions, to developments in technology and land-use and land-management strategies. In all cases, such innovations were underpinned by philosophies that placed primary value in social relationships between peoples, and between peoples and the land,

and emphasised spiritual sustainability. Such philosophies go the heart of the concept of Country.

An important aspect of Australian Indigenous innovations that needs to be kept in mind is the agency of the Ancestors and spiritual beings. For example, anthropologist Helmut Petri was informed by Aboriginal peoples of the Kimberley that the culture heroes Wódoi and Djúngun were the inventors of stone-tipped spears and spearthrowers.[26] Indeed, they pointed out that all of their 'weapons, utensils, and tools are the very same as those that the great heroes of the distant legendary past invented and used'.[27] Informed by Petri's observations and his own, anthropologist Franz-Josef Micha concluded that 'the wanderings of culture heroes, the trade, and the emergence of new religious, social, and material culture items have to be regarded as *simultaneous* processes'.[28]

These broad theoretical tenets frame our selection of 'innovations' by First Nations Australians over the past 65,000 years. We do not couch 'innovations' singularly within the language of 'new' developments or 'inventions', but as 'enhancements', to emphasise that all innovations grow out of existing social and cultural beliefs and practices. In many cases, our case-study innovations represent the earliest known examples of such practices in the world. However, we do not overplay this dimension as First Nations' history has an integrity and a significance that sit outside of global (Western) narratives of cultural development and progress.

2

ICE AGE TECHNOLOGIES OF SEA AND LAND

The extraordinary cultural and scientific knowledge of First Nations Australians is a legacy of over 65,000 years of living and observing more than 8 million square kilometres of continent. Every First Nations community had its encyclopedias of knowledge held in the heads of people but expressed physically, symbolically and spiritually across Country, like a vast living library. Today, we talk to First Nations peoples to better understand these great domains of oral knowledge embedded in Country. But what about the great knowledge of the Old People of Australia who lived not hundreds but thousands and even tens of thousands of years ago? In many cases, oral narratives provide answers to this question, but only if you know how to read and decode the traditions. In other instances, knowledge has passed beyond the oral and now resides in the ground. This *grounded knowledge* is the domain of Country, and with the help of archaeology we can investigate what the Old People left behind for us to read.

This chapter is the first of two that explore what archaeological research has revealed about the technological knowledge and innovations of the Old People from Ice Age Australia. We start with technological and skills knowledge associated with the sea and the ancient founders and settlers of Australia at least 70,000 years ago, and continue with illustrations of knowledge created to thrive in different parts of Australia between 65,000 and 12,000 years ago. The grounded knowledge focuses on stone tools as these items dominate archaeological evidence of the Old People. Chapter 3 explores spiritual and symbolic aspects of Australian Ice Age culture to help glimpse the humanity of the Old People who lived tens of thousands of years ago.

What do we mean by the Ice Age in an Australian context?

Australia's Ice Age was part of a global phenomenon where the world went through its usual 120,000-year cycle of warming for a short period followed by a long spell of cooling as the planet tilted slightly on its axis towards and then away from the sun respectively. The warm periods are known as interglacials, and we are currently in an interglacial phase; the last interglacial was around 120,000 years ago. During the glacial period, also known as the Ice Age, ice sheets of the north and south poles expanded dramatically, as did huge ice sheets (some well over 1 kilometre thick) in the Northern Hemisphere and glaciers across many continents.

Australia was cooler than today (by 5–6°C during the peak cold period known as the Last Glacial Maximum, around 20,000 years ago) and windier and drier, and like the rest of the world, sea levels dropped by 120–130 metres to expose continental shelves. Such exposure joined mainland Australia to New Guinea in the north and Tasmania in the south to form a super-island continent known by archaeologists as Sahul. As such, many sections of what we call the Australian coastline today were located tens and even hundreds of kilometres inland during the Ice Age (depending on the width of the continental shelf). It is during the middle of the Ice Age that the earliest known archaeological evidence of Aboriginal Australians appears, with the 65,000-year site of Madjedbebe in Arnhem Land.[1]

In much of the archaeological literature, the Ice Age is referred to more broadly as the Pleistocene, a geological epoch that included many Ice Ages over the past 2 million years. The Pleistocene and last Ice Age ended around 12,000 years ago, when the world entered into a warmer phase (i.e. interglacial) referred to as the Holocene. Thus, the period of Ice Age Australia associated with First Nations history, as identified by archaeological evidence, takes in the period from at least 65,000 years ago through to the end of the Ice Age around 12,000 years ago.

Before commencing our discussion of Ice Age beginnings, we note that for many Aboriginal people the question of arriving in Australia and its timing is a moot point as they say they have been here forever. For others, 65,000 years is acceptable as it is equally unimaginable and, indeed, for practical purposes forever.

FOUNDING NAVIGATORS AND SEA VOYAGERS

The technological complexity of Ice Age First Nations Australians starts big. Archaeological understandings of the arrival of the ancestors of modern Aboriginal peoples to Australia at least 65,000 years ago necessitate sea crossings using ocean-going boats from Timor and/or adjacent islands of eastern Indonesia. Even at lower sea levels during the Ice Age, the super-island continent of Sahul was never joined to Timor-Leste and Indonesia. These voyages to Sahul are the earliest known examples of open-sea voyaging by people anywhere in the world, and indicate use of sophisticated watercraft capable of successful voyaging over the horizon and across over 100 kilometres of sea. Yet these vessels are only in our imagination as no physical (archaeological) evidence exists for such technology beyond its profound legacy of First Nations Australia.

As will be shown below, current archaeological understandings of the settlement of Australia have come a long way since our undergraduate university days in the 1980s, when the prevailing minimalist view was that there 'was nothing deliberate or planned about the first settlement of Australia'. Settlement was seen as 'accidental', possibly involving only 'a single small water-craft' holding no more than 'a few castaways' represented by 'a large family – a man, two or three women and some male and female juveniles'.[2] Zoologist John Calaby even proposed sardonically that a single castaway in the form of a 'young pregnant female' clinging to a fallen tree was 'sufficient' for a 'founding population'.[3]

In the 1970s, American biological anthropologist Joseph Birdsell identified two likely migration routes to Sahul: a southern route voyaging from Timor across to the exposed north-west continental shelf of Australia, and a northern route voyaging from the north-east islands of Indonesia to the exposed western continental shelf of New Guinea.[4]

If we assume the ancient founders arrived around 70,000 years ago – that is, a littler earlier than the oldest known dated site in Australia of 65,000 years ago – then sea levels were around 75 metres lower than they are now. Following sea level and bathymetric (sea floor) modelling by economic historian Noel Butlin, more sophisticated modelling by three research teams led by archaeologists Shimona Kealy and Kasih Norman and palaeoenvironmentalist Michael Bird has better delineated how lowered sea levels exposed much of the continental shelf. This lessened the current 500-kilometre sea gap between mainland Australia and Timor to just over 200 kilometres, but remarkably also exposed a vast intervening island archipelago.[5] (Today, this spectral archipelago is again beneath the waves and no longer visible, having succumbed to subsequent sea level rise.) Such exposures allowed the ancient founders to island-hop to Australia. At sea level, the intervening islands would not have been visible from the nearest departure point of Timor; they would have come into view only after voyaging around 80–100 kilometres out to sea and over the horizon. However, from an elevated vantage point it would have been possible to see the intervening islands, and from those islands to see the distant Australian coastline (now submerged).

Wind and currents between Indonesia/Timor and Australia lowered the chances of successful random drift voyaging to almost zero per cent. Bird and colleagues conclude that the chances of successful peopling of Sahul 'by accident' is 'implausible'. Successful landfall in Australia from Indonesia/Timor required purposeful voyaging and navigational skills.

What can archaeological theorising tell us about the potential numbers and skill sets of these founding voyagers and the form and associated technology of their vessels? Demographic modelling indicates that the chances of a small group of people surviving to successfully lay the foundations for the settlement of Australia is next to zero. Complex computer modelling of population dynamics by Corey Bradshaw and colleagues reveals that a minimum of 1300 male and female founders over a relatively short period of 'several centuries' was required for successful settlement of Australia.[6] Multiple voyages by small groups spread over hundreds of years does not circumvent the problem of survivorship.

Modern DNA evidence from Aboriginal Australians (based on hair samples of the past 100 years) by Ray Tobler and colleagues reveals a confined settlement period by numerous individuals displaying genetic diversity at least 50,000 years ago, and subsequent 'genetic isolation' with no evidence for DNA additions such as would be expected from subsequent migration events.[7] The implications of over 1000 ancient founders in a relatively short amount of time discount minimalist notions of accidental voyagers and castaways. The initial settlement of Australia by the ancestors of today's First Nations peoples was planned, purposeful and, therefore, intentional.

It was a process and not an event. It is equally doubtful that the ancient founders were sailing off into the unknown. Planned and purposeful migration indicates knowledge of the viability of the destination for settlement.

Knowledge of the viability of Australia as a migration destination was likely based on two scenarios: first, deep ecological knowledge of distant vegetated lands associated with understanding the movements of migrating birds and reading the signs of bushfires and the presence of smoke on the distant horizon; and second, deep ecological knowledge of the sea to allow successful scouting and assessment trips. Such trips required two-way voyaging and the capacity to get to Australia and then, most importantly, to successfully return to Indonesia/Timor with the good news.

Intentional migration to Australia at least 70,000 years ago indicates that it was undertaken by maritime peoples with boats and high-level seafaring and navigational skills. Landlubbers or estuary dwellers would have possessed neither the boating technology nor the skills to voyage successfully over the horizon. While the form of boats made and used by these ancient founders is unknown, clearly they were capable of open-sea voyaging and could hold numerous people, both men and women and probably children.[8] We should not underestimate the boat-building skills of these ancient founders. Successful migration to Australia demonstrates unequivocally that they were able to successfully build open-sea voyaging vessels. Such ability should not be a surprise: the ancient founders were modern people with the same intellectual capacities as other contemporary peoples of the world. As archaeologists Jim Allen and Jim O'Connell cogently point out in this regard, 'inventiveness is one hallmark of modern humans'.[9]

Were the very first Australians also marine subsistence specialists? Archaeologist Sue O'Connor and colleagues note that archaeological evidence from excavations at cave and rock-shelter sites on islands adjacent to Sahul reveal that the arrival of modern people (*Homo sapiens*) heralds the 'innovation' of marine subsistence specialisation (e.g. use of marine shellfish and fish as food) in Indonesia/Timor.[10] Interestingly, it is these coastal peoples who made the first journeys to Australia, an insight that matches up with the theory that the ancient founders had sophisticated maritime voyaging skills.

Why these ancient founders decided to migrate to and settle on the ancient shoreline of north-west Australia is an unanswered question open to contemplation and speculation. Butlin was probably on the money when he suggested that the motives were likely a complex mix of 'push' and 'pull' factors.[11] We do know that it was the first great maritime migration in human history. It's time to lock in Butlin's contention that the ancestors of First Nations Australians can 'claim the title of the world's first mariners'.[12] We also know that subsequent generations of these First Peoples discovered that they had an entire continent of extraordinary environmental diversity with endless opportunities to explore, settle and thrive.

NEW STONE-TOOL TECHNOLOGIES

Archaeological evidence of Ice Age technologies used by the Old People of the Australian continent is limited greatly by issues of preservation. Although most Ice Age tools were organic and made of wood, similar to recent centuries, time and preservation have conspired to bias our insights largely to stone tools. In rare cases, Ice Age bone tools have preserved. Very rarely do we glimpse the wooden dimensions of Ice Age technologies: the two key examples both come from the very end of the Ice Age. First, numerous Gwion Gwion painted figures from the Kimberley region of north-west Western Australia dating to around 12,000–13,000 years ago (i.e. the end of the Ice Age) hold boomerangs, spears, woven bags and digging sticks. Second, more direct evidence comes from Wyrie Swamp in South Australia, where archaeologist Roger Luebbers excavated boomerangs, digging sticks and a spear preserved in peaty deposits dating to between 10,000 and 12,000 years ago.[13]

Despite issues associated with the preservation of Ice Age organic tools in Australia, stone tools provide extraordinary scope to understand the innovative practices of the Old People. Mastery of stone and fracture mechanics for the creation of stone tools is one of the great innovation hallmarks of humanity. The oldest known stone tools come from East Africa and date to 3.3 million years ago, with the earliest stone tools associated with *Homo* (humans) dated to 2.6 million years. Within Australia, stone tools mark the oldest and deepest levels of the Madjedbebe rock-shelter in Arnhem Land, dated to 60,000–65,000 years ago.[14]

Early archaeological views on the nature of stone-tool technologies employed by Ice Age Australians emphasised simplicity and pan-continental homogeneity, coming under the bland label of 'the Australian core tool and scraper tradition'.[15] Dramatic increases in the number of investigated Ice Age sites coupled with new and more sophisticated techniques of stone-tool analysis over the past thirty years have seen understandings of Australian Ice Age technologies change dramatically. It is now clear that sophistication, complexity and regional diversity underpin stone-tool use across Ice Age Australia between 12,000 and 65,000 years ago. Three dimensions of Ice Age stone-tool technologies are described below to illustrate this sophistication, complexity and diversity: ground-edge axes, thumbnail scrapers, and heat treatment of raw materials.

Ground-edge axes

Ground-edge axes are one of the most easily recognised stone-tool types. Talk to most farmers in Australia and they will show you ground-edge axes they've found on their property or at least tell of similar tools located by their neighbours. Most ground-edge axes in Australia are palm size and display a distinctive oval form that has been chipped into shape, followed up by grinding and smoothing of one end, which is the cutting/chopping edge. Grinding was performed with the aid of a grindstone, usually of an abrasive material such as sandstone, and the application of water. Some axe grindstones were clearly handheld and portable; in other cases people used bedrock at the edges of creeks, recognisable today by linear depressions created by years of grinding activity. A low proportion of Aboriginal axes are ground smooth over most of their surface but none are completely smoothed, as seen with many Melanesian and Polynesian stone axes. Comprehensive grinding and polishing is not done for functional reasons: it is undertaken for aesthetic and symbolic reasons. Raw materials include igneous rocks, such as basalts, and metamorphic rocks, such as greenstone (metamorphosed basalt), hornfels (metamorphosed fine-grained sedimentary rocks) and quartzite (metamorphosed sandstone). Ethnographic observations reveal that Aboriginal stone axes included a wooden handle held in place with plant-fibre packing and plant resins.[16]

Aboriginal stone axes were multipurpose tools used for a wide range of tasks, including chopping wood and bark to manufacture shelters, canoes, shields and spearthrowers; chopping into trees to access possums, honey and grubs; butchering of animals and skinning; and use of the flattened 'butt' end of the axe head for pounding bark for twine, and pounding nuts, kernels and roots for food.

Archaeologists Val Attenbrow and Nina Kononenko undertook a comprehensive study of the use-wear patterns on a sample of fifty-one ground-edge axes from the Central Coast region of New South Wales held by the Australian Museum in Sydney.[17] The majority of axes revealed use-wear evidence of woodchopping and handle hafting (as expected), followed by plant-food processing, polishing/abrading/breaking bones/shells, flaking stone to make other tools, skin scraping, and pounding ochre pigment. Although most ethnographic records indicate use of axes by men, Attenbrow and Kononenko posit that plant-food processing functions strongly suggest axes were also used by women.[18] In most cases they were prestigious status objects with deep symbolic and cosmological significance, often obtained through trade (see Chapter 4).[19]

Anthropological understandings of the origin and antiquity of Aboriginal ground-edge axes got off to a shaky start. In the 19th century, European anthropologists and archaeologists pronounced ground-edge axes a hallmark of the Neolithic, the period when Europeans began growing crops and clearing forests to make fields. Flaked-stone tools were from the earlier Palaeolithic period. This sequence created a conundrum for late-19th-century European scholars because Aboriginal Australians, considered living fossils of the Palaeolithic, also made and used ground-edge axes. In short, how could a so-called single group of people possess stone-tool technologies belonging to two entirely different time periods? The stakes were high because such an apparent mixture went against theories of the time that emphasised racial stereotypes and cultural hierarchies. For founding Oxford anthropologist Edward B Tylor, the answer to this anthropological conundrum was straightforward: Aboriginal Australians possessed a Palaeolithic technology, but the Neolithic technology was a recent influence and an add-on from Pacific agricultural peoples who were, so the story goes, at a more advanced (Neolithic) stage of cultural development.[20] Remarkably, colonialist beliefs that ground-edge axes were a recent external addition to Aboriginal culture from culturally advanced Pacific neighbours held sway for much of the 20th century.[21]

Until the early 1960s, the earliest archaeological evidence for ground-edge axes in the world came from Neolithic Europe, including the United Kingdom, dating to around 6000–7000 years ago.[22] Then, in the first half of the 1960s, came the bombshell discovery that Australian Aboriginal ground-edge axes dated to at least 20,000 years ago. This revelation was a result of excavations undertaken by archaeologist Carmel Schrire at three rock-shelter sites in the Kakadu region of Arnhem Land.[23] Over the past forty years, archaeological research has revealed the manufacture of ground-edge axes in Europe during the Mesolithic period (i.e. immediately before the Neolithic) dating to 8000–9000 years ago.[24] At the same time, ground-edge axes dating back over 20,000 years have been documented in eastern Asia, with the earliest known examples being from Japan and dating to 32,000–36,000 years ago.[25]

With excavations at Madjedbebe in Arnhem Land, Australia was the site of yet another astonishing finding about the origins and antiquity of global ground-edge axe usage. Archaeologist Chris Clarkson and colleagues recovered numerous ground-edge axes among thousands of smaller flaked-stone artefacts in the lower levels of the site dating to between 55,000 and 65,000 years ago.[26] Clarkson, a stone-tool expert, employed a wide range of techniques to show that the axes were actually deposited in these ancient layers, as opposed to falling down cracks in the sediment from higher-up and younger levels of the site. Some of the Madjedbebe axes feature a groove around the upper sections that was made to accommodate a handle. Similar grooves were identified by Schrire in the axes she excavated from Arnhem Land in the 1960s. The ancient Arnhem Land axes are the oldest known examples in the world of ground-edge axes and of people hafting handles onto tools.

The Madjedbebe axes are not an isolated find: similar ground-edge axes have been found at other Ice Age sites dating to 30,000 to 50,000 years ago across far northern Australia, from the Kimberley (Western Australia) through to Cape York Peninsula (Queensland).[27] Curiously, Ice Age axes have yet to be documented for the southern half of Australia, with the earliest examples no more than 4000 years old.[28] The Ice Age tradition of ground-edge axe manufacture in Australia was highly localised and limited to the tropical regions of the far northern parts of the continent.

Archaeologists Anne Ford and Peter Hiscock point out that ancient Australian Ice Age axes highlight the 'innovation and flexibility' of Australia's earliest settlers and 'mirrors their success in adapting to a wide range of habitats soon after their first arrival'.[29] One question that remains unanswered is whether ground-edge axe technology arrived in Australia with the ancient founders or was developed after their arrival. Unfortunately, no archaeological sites in Indonesia/Timor of similar age to Madjedbebe have been excavated to see if they contain ground-edge axes. It remains entirely possible that ground-edge axes were an innovation of the earliest settlers of Australia.

Thumbnail scrapers

Towards the final stages of the Ice Age in Australia, Aboriginal Tasmanians rapidly expanded their manufacture and use of small scrapers known by archaeologists as thumbnail scrapers. The best-documented examples were recovered during excavations of a number of cave and rock-shelter sites in the remote south-west of

Tasmania during the late 1980s and early 90s.[30] Today the region is blanketed in dense temperate rainforest, but during the late Ice Age between 12,000 and 40,000 years ago when these sites were variously occupied, it was mostly grasslands in sight of glaciers.[31] Thumbnail scrapers are small (<2.5-centimetre) tools, discoid or semi-discoid in outline, with a curved working edge created by micro-chipping (termed 'retouching' by archaeologists) at one end and sometimes further retouching along the sides, which may have been to accommodate a handle haft. In some cases, the tools have been resharpened several times as their edges blunted through use.[32]

Although south-west Tasmanian thumbnail scrapers date to 33,000 years ago, the majority have been recovered from archaeological deposits dating to after 20,000 years ago.[33] They are made from high-quality local raw materials such as chert, crystal quartz and milky (white) quartz. Some are made from tektites known as Darwin glass, comprising pebbles of glass that formed when a meteorite struck south-west Tasmania around 800,000 years ago. The impact, marked by what is known today as the 1.2-kilometre-wide Darwin Crater, located south-east of Queenstown in central-western Tasmania, melted local sediments and rocks to form glassy tektites that were strewn across 400 square kilometres of landscape in a westerly direction from the crater.[34] Ice Age Tasmanians would have found tektites on the ground surface and eroding from surface sediments as they traversed their Country. Selection was limited to larger tektites that could be flaked to form small scraping tools.

Use-wear analyses of Ice Age thumbnail scrapers from Kutikina Cave by archaeologists Tom Loy and Richard Fullagar revealed that the tools were multifunctional, with uses including butchering and the scraping of wood, bone and skin. In contrast to Loy's analysis, Fullagar found signs of rubbing consistent with insertion of thumbnail scrapers into wooden handles.[35]

Ice Age thumbnail scrapers have also been documented from an Ice Age site on Hunter Island in Bass Strait (then Hunter Hill on the Bassian Plain) and a probable Ice Age archaeological site at Green Gully near Melbourne in central-southern Victoria.[36] Although thumbnail scrapers do occur across other parts of Australia over the past 5000 years, during the Ice Age their use seems to be restricted to south-eastern Australia, including Tasmania.

Obviously, thumbnail scrapers and ground-edge axes demonstrate that technical advances by Aboriginal peoples in Ice Age Australia were locally focused, with regional groups innovating regional answers to technological needs.[37] It is also possible that such regional traditions were expressions of locally distinct Ice Age cultures.

Heat treatment of stone

A critical dimension to the mastery of stone for tool manufacture is selection of particular types of raw materials that are conducive to flaking and shaping into desired tool shapes and forms. Most types of rock are not suitable for making stone tools. As such, stone-tool technology requires comprehensive geological knowledge of the quality of rock outcrops across large areas of landscape. Some types of stone, such as obsidian (volcanic glass that is almost pure silica), are perfectly suited to tool manufacture because they can be easily chipped into shape and create razor-sharp cutting edges. However, this sharpness comes at a cost as obsidian is brittle and less suited to working and chiselling wood. In this case, stone-tool producers select different raw materials that are conducive to flaking and also create strong edges that will withstand the impacts of working hardwoods. Although Australia has only a couple of sources of obsidian, all poor quality (in contrast to eastern Papua New Guinea, where large outcrops of obsidian occur), the continent is blessed with a plethora of high-quality stones that are perfectly suited to the manufacture of flaked cutting and scraping tools. All of these high-quality stone types are very high in silica; they include chert, silcrete and quartz.

Remarkably, ancient peoples in various parts of the world knew that the quality of raw materials for manufacture of stone tools could be increased by artificially changing the microcrystalline structure of the raw materials through careful and highly controlled application of indirect heat from fire (ideally 400–600°C).[38] The earliest known archaeological evidence for mastery of the heat treatment of tool stone comes from South Africa, dating to 130,000 years ago.[39] These silcrete artefacts show visual signs of heat treatment, such as colour change and smooth flaked surfaces (pre-heat-treated flake surfaces tend to have a rough texture). That is, a block of stone raw material of the necessary size and thickness, often shaped by flaking, is placed into previously prepared fire-heated sediments – such as at the margins of a fireplace or within a ground oven – for many hours. After removal and cooling, the flaking of the block takes place, allowing a visual contrast to be made between the pre- and post-heat-treatment flaked surfaces. The stone-tool makers would have noticed an obvious change in the way the heat-treated block flakes.

Put simply, flaking is easier and more predictable in its outcome because heat treatment makes the stone physically harder and more homogenous. Such microscopic changes allow the impact force to penetrate the stone unimpeded, creating a smooth flake surface. As a result, stone-tool makers can more easily create the exact-shaped tools they need.

What is the earliest evidence for heat treatment of tool stone by First Nations Australians? Research by archaeologists Patrick Schmidt and Peter Hiscock has revealed heat-treated stone artefacts made of silcrete at Lake Mungo and a range of other Ice Age sites in the broader Willandra Lakes Region of western New South Wales dating variously to between 15,000 and 42,000 years ago.[40] The heat-treated artefacts were recovered from camp sites eroding out of dunes along the edges of ancient lakes, with boulder and cobble outcrops of silcrete easily accessible nearby.[41] Examination of silcrete outcrops has revealed extensive evidence of quarrying activity and initial flaking of extracted blocks of silcrete into usable cores for transportation to nearby camp sites for follow-up tool manufacture and use. Just over 60 per cent of the 509 silcrete artefacts examined from Lake Mungo showed signs of heat treatment. Schmidt and Hiscock note that this level of usage indicates that Ice Age peoples of the Willandra Lakes Region between 17,000 and 42,000 years ago had 'fully mastered and wholly integrated' heat treatment of silcrete artefacts as a 'regular and systematic' part of their stone-tool technology.

Interestingly, more recent silcrete artefacts in the Willandra Lakes Region show less evidence of heat treatment. For example, the proportion of heat-treated silcrete artefacts at the 3800-year-old

Lake Garnpung camp site is 11 per cent; it is 9 per cent at the 600–900-year-old Lake Mungo site.[42]

It is likely that the ancestors of peoples who first settled the Willandra Lakes Region at least 50,000 years ago arrived with knowledge of heat treatment of tool stone. The Willandra Lakes are over 3000 kilometres south-east of the closest entry point to Australia, at the edge of the then-exposed continental shelf of the Kimberley. As multiple generations of these ancestors explored and settled huge areas of Australia, they would have had multiple opportunities to refine heat-treatment processes to the needs of Australian tool stones. We can imagine that the earliest Old People of the Willandra Lakes Region quickly fine-tuned their technological knowledge, skills and science to the specific heat-treatment requirements of local silcretes.

Yet peoples of the semi-arid Willandra Lakes Region were not the only Ice Age Australians employing heat-treatment technology. Schmidt and Hiscock have also investigated the presence of heat treatment of silcrete artefacts in two other, vastly different contexts: arid Central Australia and temperate coastal south-eastern Australia. The first example from Central Australian rock-shelters comprised 296 artefacts from Puritjarra dating to the past 22,000 years, and 118 artefacts from Kulpi Mara dating to the past 34,000 years.[43] Both sites averaged highest proportions of heat treatment (67–74 per cent) in Ice Age levels and lowest proportions (48–54 per cent) in layers dating to the past 8000 years. For arid Central Australia, the apparent trend mirrors that of the semi-arid Willandra Lakes of decreasing heat treatment through time. The second example comprised nearly 800 silcrete artefacts from different levels of the 29,000-year cultural sequence at Burrill Lake rock-shelter, located south of Wollongong in the Sydney Basin.[44] Results reveal increasing proportions of heat-treated artefacts through time, ranging from 50 per cent in the oldest Ice Age layers (29,000 years ago) through to 76 per cent in the most recent layers (dating to the past 1300 years). This overall trend of increasing heat treatment through time has also been documented by Schmidt and Hiscock at other sites in the Sydney Basin. Possible reasons advanced for the trend are enforcement of tribal boundaries that limited access to high-quality tool stone quarries and/or an increasing need for higher-quality raw materials to manufacture more finely made tools requiring high-precision micro-flaking of edges.

Increasing use of heat treatment through time in the Sydney region of temperate south-eastern Australia contrasts with a moderate decrease in heat treatment through time in arid Central Australia and a major decrease through time in the semi-arid Willandra Lakes Region. Schmidt and Hiscock rightly point out that these regional differences in the long-term history of heat treatment of stone artefacts indicate that no single pan-continental explanation exists for heat treatment by First Nations Australians. Explanations must be regionally based and focus on local cultural differences and requirements, and on innovative responses to local geological conditions. In the next chapter, we meet two people from Lake Mungo who died 40,000 years ago and who would have known about heat treatment of tool stone.

3

ICE AGE BURIALS AND FUNERALS

The Old People of Ice Age Australia developed some of the world's earliest known expressions of spirituality and symbolism and what it means to be human. This insight will not surprise many First Nations Australians who have a deep sense of their spiritual connection to Country and Ancestors that is timeless. What can archaeological research add to these understandings?

Australia has revealed ancient burials that date back tens of thousands of years. Among the millions of stone tools that dominate physical (archaeological) evidence of Ice Age Australians, very occasionally one of the Old People makes their grave evident. For current generations of First Nations people, the impact of being witness to the appearance of an ancestral grave is immediate and profound. Here one of the Old People has come through time to meet up with their modern descendants. Often the message they bring is intimate and sacred and only for the ears of First Nations people. Yet in many cases, First Nations Custodians of these remains know that some messages need decoding with the curated assistance of archaeologists and physical anthropologists, who might be invited to undertake scientific studies guided by strict cultural protocols.

In the past, First Nations peoples had no say in who excavated their ancestors and who took away their remains for study. Abhorrent colonialist science with its racist agendas continues to haunt First Nations peoples in Australia and in other colonial contexts such as North America and Africa. Repatriation of ancestral remains from museums and other research institutions to descendant communities can help heal wounds of the past. Today in Australia, no archaeologist or physical anthropologist would excavate or study ancestral remains unless local Traditional Owners sanctioned and participated in the research.

In this chapter we provide examples of Ice Age spirituality and symbolism in the form of graves and associated ancestral remains and grave objects, as revealed by detailed scientific analysis. We focus on the 40,000-year-old cremation of Mungo Lady and the similarly aged ochred burial of Mungo Man from western New South Wales, and the 20,000-year-old cemetery at Kow Swamp in central-northern Victoria. In both cases we are dealing with the rarest and what in certain cases are regarded as the most significant and sacred archaeological sites documented on the Australian continent. These remains are considered to be innovative examples of mortuary practices (burials and funerals) that are either the earliest known examples in the world or at least *among* the earliest known examples. In all cases, these burials also have significance outside of Australia as they help other peoples around the world understand the bigger picture of the cultural development of our species.

Before we discuss the Lake Mungo and Kow Swamp people, it is important to know that most archaeologists in Australia who have had the rare experience of excavating the grave of one of the Old People have been profoundly touched by the sanctity of the remains and humbled by the experience.

IAN J McNIVEN

During the 1980s, I undertook my PhD on how Aboriginal peoples lived across the Cooloola region of coastal south-east Queensland. My focus was archaeological evidence such as shell middens and stone artefacts. I examined old camp sites along the coast and small and ephemeral camp sites comprising a single meal of shellfish and the occasional stone tool located up to 12 kilometres inland. After a few years, I felt like I was starting to understand how the Old People lived at Cooloola and that the earliest evidence of this coastal lifestyle extended back at least 5500 years. I discussed my research with the local Butchulla and Kabi Kabi communities, and always thought about the Old People who had eaten the meals and made and used the stone tools I touched and studied. Then, in 1989, all that changed when I had the rare and extraordinary opportunity to meet two of the Old People.

In April 1989, the Double Island Point lighthouse keeper informed local National Parks rangers that wind erosion had exposed what appeared to be a skull associated with an Aboriginal burial in a nearby dune field. I was contacted, and after a visit to the site confirmed that it was indeed an Aboriginal burial. The burial was highly vulnerable to interference from tourists, and it was clear that placing sand back over it would only be a short-term measure as the sand would soon blow away and expose more of the body. Following discussions with the Queensland Government heritage office and senior members of the local Butchulla community, it was decided that the best course of action was to excavate the burial with an agreement that the remains would be reburied nearby in a stable, vegetated dune after potential scientific description.[1]

Subsequent excavation of the burial by archaeologist Bryce Barker and me in late April revealed the highly weathered remains of a middle-aged man who had been buried in a semi-flexed position lying on his back with his knees pointing down, his feet brought up towards his buttocks, and his head tilted to the left. After exposing the remains and plotting each bone on graph paper, we carefully wrapped each bone in tissue paper before placing it into a labelled plastic bag for protection. It wasn't feasible to plot and collect minute fragments of bone. My fieldnotes for 22 April read: 'It is probably a nice symbolic gesture that a few fragments stay there anyway!' As per legal requirements, I delivered the burial remains to the Queensland Government heritage office in Brisbane on 24 April.

When visiting the burial location a few weeks later in June, I noticed that continued wind erosion in the area had exposed another Aboriginal skull with attached jaw, indicating that the rest of the body was likely to extend under the sand. Again, it was considered appropriate by the government heritage office and the Butchulla people for me to excavate the burial for protection, potential analysis and subsequent reburial. I was assisted by archaeologists Bryce Barker, Su Davies and Kathy Frankland from the University of Queensland.

We excavated the second burial in early June and agreed that it was an extremely moving and humbling experience. Unlike the first burial, the second was of a young woman whose remains were extremely well preserved. I will never forget the image of her fully exposed skeleton resting within her oval-shaped grave. Like the first burial, her relatives had buried her in a semi-flexed position with knees pointing down and feet brought up to touch her buttocks. Her hands were placed in her lap and her head was resting to the right. We all agreed that the young woman looked like she was asleep. Remarkably, we noticed a small brass safety pin resting on her right chest. This pin dated the young woman to after European invasion and meant she was probably one of the last Aboriginal people of 19th-century Cooloola to be buried in a traditional way on her Country. We believe the brass pin held in place an article of clothing, possibly a shawl. As with the first burial, we remained in constant contact with senior members of the Butchulla community throughout the excavation. The young woman's remains were wrapped carefully as for the first burial and taken to the government heritage office in Brisbane on 15 June.

Following advice from Butchulla Elder Aunty Olga Miller, the Kabi Kabi people took possession of the burials. They subsequently informed me that the man and young woman had been reburied at an undisclosed location near Double Island Point, where they continue to rest in peace.

LYNETTE RUSSELL

In the early 1990s, after completing my undergraduate degree in Archaeology, I was briefly employed at Aboriginal Affairs Victoria's Heritage Services Branch. My role was to undertake Indigenous community engagement and assist the senior state archaeologist on various field trips. One of my earliest trips was to north-western Victoria, where we were to monitor a number of burials recorded in the 1970s by Robert Blackwood and Ken Simpson. The burials had been stabilised using jute matting, and we were to record the impact of wind and water erosion and see if there was any evidence of vandalism or other destruction, such as might be caused by feral pigs. I had no idea at the time, but this trip was to be one of the formative experiences of my life.

The burials were dug into the fine red sand and marked with aluminium 'dog tags' that were fixed to metal tent pegs. Most of the burials were well-covered, and the jute matting and wind-deposited sand protected them. One burial, which a physical anthropologist had thought to be an elderly woman, had been exposed. Her grave had only recently been uncovered by wind. The soil surrounding her head was stained, probably with vegetation such as rootlets, in such a way that I thought (or rather felt) I could make out a face. Her mouth had fallen open as the soil shifted and moved. She looked to me as if she was calling out something important across the millennia. There was something utterly recognisable and completely human about her. It felt surprisingly intimate. The immediacy and the fragility struck me and, although knowing full well that I had no relationship to her, I felt a deep sense of connection.

KOW SWAMP CEMETERY

The Kow Swamp Aboriginal cemetery site, located in northern-central Victoria just south of the Murray River, was identified by physical anthropologist Alan Thorne in February 1968. Its location followed detective work by Thorne and Alan West of the National Museum of Victoria (now Melbourne Museum), who tracked down the source of an unregistered box of Aboriginal skeletal remains with calcium carbonate encrustation that had been deposited in the museum by Bendigo police in 1962. The remains had been exposed accidentally by construction of an irrigation channel on the eastern margin of Kow Swamp. A series of excavations at the site by Thorne and West during 1968 recovered more remains of the person in the museum box, and the presence of other burials. Assisted by geomorphologist Phil Macumber, Thorne led large-scale excavations within the lunette (dune) and underlying silts along the eastern margin of Kow Swamp between 1970 and 1973 to better understand the form, scale and age of burials at the site.[2] The local Yorta Yorta community neither sanctioned the excavations nor participated in the fieldwork.

The remains of more than forty people were recovered from three different excavation locations at Kow Swamp between 1969 and 1973.[3] Most of these people date to the Ice Age and are in a poor state of preservation, with many sections of bone completely dissolved away. As such, only twenty-two people (designated by Thorne as KS1 to KS22 but referred to here as Persons 1 to 22) have been formally identified and described.[4] The remains of nineteen people were found in a 2.5-hectare area with a concentration of eight oval-shaped graves spaced 1–4 metres apart, containing the remains of eleven people confined to an 11-metre × 5-metre area. People had been buried in a horizontal flexed position with the ankles brought up close to the buttocks and in some cases the knees in front of the chest. The exceptions are Person 1 (horizontal extended position) and Person 5 (vertical crouched position). More than half (15) of all the people buried at Kow Swamp are adults, followed by juveniles (4), infants (2) and an adolescent (1). Of the thirteen individuals who could be sexed and aged (based on preservation of the pelvis and teeth), ten were male (all adults) and three were female (juvenile to young adult).

All graves but two contained a single person. The multi-person graves contained a primary grave with additional remains. The grave with Person 13 included Person 21 (partial remains of an infant) and Person 22 (partial remains of an adult). The grave with Person 16 included an extra talus (ankle bone). Except for the infant, Thorne suggests that the additional remains may have been mortuary relics of individuals that had been carried around for some time prior to interment.

Four people were buried with grave objects of deep ritual and symbolic significance that had withstood the passage of time. Person 5 (vertical crouched position, adult) was associated with freshwater mussel shells, stone tools made from quartz, and fragments of ochre. Person 14 (horizontal flexed, adult male) had whole and fragmented mussel shells sprinkled over his body and a cluster of ten complete mussel shells positioned under the left side of his head. Person 16 (horizontal flexed, adult female) had a remarkable row of kangaroo upper-jaw incisors placed across the upper part of her skull, clearly the remains of a headband, and powdered red ochre sprinkled across the upper half of her body. Person 17 (horizontal flexed, adult male) was buried with quartz artefacts and fragments of mussel shell. Thorne focused his attention on the biology of the human remains, not the extraordinary grave objects. At that time, Western scientific eyes were on the evolutionary significance of the Kow Swamp remains, not their social and cultural importance.

Radiocarbon dating of mollusc-shell grave objects and human bone by Thorne and Macumber indicated that the people were buried at Kow Swamp sometime between 10,000 and 15,000 years ago.[5] Subsequent dating of sand grains within sediments using the OSL (optically stimulated luminescence) technique by archaeologist Tim Stone and geochronologist Matt Cupper suggest the burials are much older and date to between 19,000 and 22,000 years ago.[6] This age span takes in the peak cold and dry period of the Ice Age known as the Last Glacial Maximum, which centres on 20,000 years ago.

Thorne interpreted the sloping forehead and robust features of the Kow Swamp people as inherited ancestral 'features' of pre-modern human *Homo erectus* in Indonesia.[7] Subsequent reanalysis of the remains by physical anthropologist Peter Brown demonstrated that the distinctive skull shapes of some of the Kow Swamp people was due to the cultural practice of head-binding[8] – that is, pressure was applied by mothers to the malleable skulls of their infants to create more oblong-shaped heads. Head-binding is a rare cultural practice that was undertaken by a number of Indigenous societies (e.g. in Torres Strait, New Guinea and North America) until the late 19th century; archaeological evidence indicates more widespread application of head-binding in South America. The Kow Swamp people are the oldest known examples of head-binding and represent the earliest known example of people deliberately changing the shape of their skeletal bodies for cultural reasons. Although body reshaping and modification are common practices these days (e.g. plastic surgery, dental enhancement), current archaeological evidence indicates that the practice was an innovation of Ice Age Australians.

The innovations of the Kow Swamp people also include the development of cemeteries for the dead. Archaeologist Josephine Flood notes that Kow Swamp represents 'the largest single population of the late Pleistocene epoch found in one locality anywhere in the world'.[9] More significant is the concentration of burials, which is consistent with a formal burial ground or cemetery. Kow Swamp is currently the oldest known cemetery in the world. Physical anthropologist Colin Pardoe points out that the highly visible mortuary practice of burying the dead in cemeteries was commonplace along the Murray River over the past 6000 years, and was an expression of corporate group identity and associated territorial rights and responsibilities, deep ancestral connection and land ownership.[10] Kow Swamp signals the first evidence of this cemetery tradition, with Pardoe adding that the practice of head-binding may have been an expression of corporate group identity and the desire to look different to your neighbours.[11] The new dating of the Kow Swamp cemetery reveals that such corporate group identity and land ownership extends back to the Ice Age to at least 20,000 years. It is against this extraordinary backdrop of long-term ownership and property rights that we need to comprehend the moral legitimacy of the Victorian Government's sanctioning of the return of the Kow Swamp burials (ancestral remains and associated grave objects) to the Yorta Yorta First Nations community for reburial in 1991.[12]

LAKE MUNGO CREMATION

The story of the 'discovery' of Mungo Lady and Mungo Man by geomorphologist Jim Bowler on 5 July 1968 and 26 February 1974 respectively as he walked alone along the Lake Mungo lunette has been told many times.[13] Today, Bowler would be the first to agree that the term 'discovery' doesn't quite convey the spiritual importance of the Mungo remains. Did he discover the remains, or did they discover him? Whatever the case, Bowler soon realised that he wasn't alone when walking the ancient shores of Lake Mungo. Traditional Owners such as Aunty Dorothy Lawson know that 'She surfaced for a reason',[14] and Mutthi Mutthi Elder Aunty

Mary Pappin noted: 'I believe that the Mungo Lady came to walk with our people to help us with our struggle and to tell the rest of the world about our cultural identity with the land.'[15] In a cross-cultural gesture, on those fateful days Mungo Lady and Mungo Man made themselves known to Jim Bowler. But their reveals set in place a complex set of circumstances where archaeologists and physical anthropologists failed to involve local Mutthi Mutthi, Paakantji and Ngyiampaa Traditional Owners in deciding the immediate fate of the remains. Deciding exactly what should have taken place was complicated by the varying wishes of the three different First Nations communities whose Country takes in Lake Mungo. What did happen with the eroding remains is well documented.

In March 1969, Mungo Lady (then referred to as Mungo I) was excavated by a research team from the Australian National University (ANU) in Canberra that included archaeologists John Mulvaney, Rhys Jones and Harry Allen.[16] The remains were handed to physical anthropologist Alan Thorne (of Kow Swamp fame), also at the ANU.[17] In the laboratory, Thorne used miniature drills and acid to remove the highly fragmented bones one by one from a matrix of carbonate that had formed around the remains over thousands of years. About 2000 bone fragments were recorded; the cranium alone was in 175 fragments. It took Thorne over six months to meticulously extract the bones and painstakingly reconstruct the skull.[18] Although only 25 per cent of the skeleton was present, he was able to work out that the bones belonged to 'a young female adult of gracile build and small stature'.[19] That is, Mungo Lady's teeth indicate that she died young, at an age of 20–25, and that she stood 150 centimetres tall.[20]

Soon after excavation, Mungo Lady was dated to 25,000–32,000 years ago based on radiocarbon dating of charcoal and shells from nearby camp sites. Since then, Bowler has refined his chronology for the Ice Age dune sequences at Lake Mungo, resulting in a new date for Mungo Lady centring on 40,000 years, within the range of 38,000–42,000 years.[21] This new dating also established that the earliest stone artefacts at Lake Mungo date to between 46,000 and 50,000 years ago. As such, Mungo Lady was not one of the first residents of Lake Mungo – but she is certainly the oldest known woman from the region and, indeed, all of Australia.

Many of Mungo Lady's bones showed signs of major burning, which led Thorne to conclude that the young woman had been cremated and her bones deliberately fragmented. Severe burning on the left side of the head suggested that she had been placed on her left side on the funeral pyre. Some of the broken bones looked like they had been reburnt, indicating that they had been removed from the funeral pyre, fragmented further, and then returned to the pyre; the bone fragments, along with some of the pyre ash (charcoal), were then gathered together and placed in a nearby small shallow grave measuring approximately 75 centimetres wide and 20 centimetres deep, within a few metres of the lake edge. It is not difficult to see the elaborate treatment of Mungo Lady's body as a complex funerary ceremony by her loved ones. As she was a young woman, it is entirely plausible that her parents were among the mourners. During the 19th century, over 1400 generations later, Aboriginal people of eastern Australia (including Tasmania) continued the tradition of cremation of loved ones.[22]

When the existence of Mungo Lady came to light in the late 1960s, archaeologists recognised that her funeral represented the earliest known example of cremation in the world. Subsequent archaeological research across the globe over the past fifty years has not changed that conclusion, or the fact that the practice of cremation is a First Nations Australian innovation. Elsewhere in the world, archaeological evidence of cremations is restricted to the past 12,000 years.[23]

LAKE MUNGO BURIAL

In contrast to the cremated remains of Mungo Lady, Mungo Man was a full-body burial. Within two days of Bowler noticing Mungo Man's skull eroding from the ancient sands of Lake Mungo, a research team from the ANU that included Alan Thorne and archaeologist Wilfred Shawcross collaborated with Bowler to excavate the remains (then referred to as Mungo III) before they succumbed to wind and rain erosion and the destructive hooves of sheep.[24] After a number of days, the excavation team exposed the fully extended skeleton of an adult male. Despite their 'extremely fragile' state, all the bones were more or less in correct anatomical position, leading Thorne to rightly conclude that Mungo Man had been buried shortly after death. The fragile state of the bones necessitated removing the body in a series of sand blocks for more careful conservation and excavation at the ANU.

A remarkable aspect of Mungo Man observed during excavation was 'pink staining' of sediments around his body. This staining stood out against the local white sands and clearly represented red ochre powder that had been sprinkled over the body before it was covered with sand. The ochre allowed delineation of the original 'shallow' grave, which measured 1.8 metres by 1.3 metres with a depth of 80–100 centimetres.[25] As such, the grave was just large enough to accommodate the man's height of 175 centimetres.[26] The position of the bones indicated that Mungo Man had been placed in the grave on his back with a slight rotation to the right, including his head tilted towards his right shoulder. His hands were placed together in his lap and his left leg rested over his right leg. Both legs were slightly bent, perhaps to help fit into the grave. His right elbow was resting on the edge of the grave, as was his left foot.

Detailed examination of Mungo Man's body by physical anthropologist Steve Webb indicated that he was around fifty years of age and suffered from severe osteoarthritis in his spine and right arm, possibly from infection.[27] Interestingly, both canine teeth in the lower jaw of Mungo Man were missing. The bone sockets of the removed canines had been partly reabsorbed, indicating tooth removal 'many years' before his death. Webb also noted that the canines must have been removed purposefully, clearly for ritual reasons similar to ritual tooth avulsion of one or two incisors of men and women for initiations known historically across many parts of Australia.[28] The surfaces of some of his molars show wear patterns in the form of 'distinct grooves' and 'striations' consistent with processing plant fibres, perhaps for the manufacture of string for fishing nets or dillybags.[29] If it was for a fishing net, then it relates to fishing in a lake that hasn't seen water for 15,000 years.

Although Mungo Man was buried in dune sands 'almost indistinguishable' from those of Mungo Lady, he was assigned a slightly older age of 28,000–32,000 years by Bowler.[30] However, his age has since been reassessed by Bowler and is now similarly considered to be 40,000 years.[31] Even after 40,000 years, archaeological and physical anthropological science can give a voice to Mungo Man and have him tell us about his life.

Mungo Man is the oldest known burial of a complete person in Australia and the rest of the New World (including the Americas). The oldest known burial of a modern person (*Homo sapiens*) comes from the cave site of Skhul in Israel, dating to 100,000–130,000 years ago.[32] Across northern Eurasia, the nearest old burials of modern people come from Kostenki and Sungir in Russia, dating to 30,000–38,000 years ago.[33] Across Eurasia, it is very likely that numerous excavated Neanderthal bodies (*Homo neanderthalensis*) also represent intentional burials with some sort of symbolic associations between around 40,000 and 70,000 years ago, but the issue remains contentious among archaeologists.[34]

Neanderthal burials, however, are not associated with grave goods or other ritual accompaniments; the earliest unequivocal evidence in the world for a modern human burial with ritual objects is an infant burial with a perforated cone-shell adornment in South Africa dating to 74,000 years ago.[35] By this count, Mungo Man, with his red ochre adornment and dating to 40,000 years ago, is currently the oldest known burial of a person associated with ritual objects outside of Africa.[36] Burial of the dead with ritual symbolism is a key defining cultural practice of humans, and Mungo Man reveals that First Nations Australians were front and centre in this innovative development.

Mungo Man also appears to be the earliest known evidence for tooth avulsion. Given that such selective tooth removal is usually undertaken ritually in association with initiation and rite-of-passage ceremonies and the ceremonial advancement to adulthood, Mungo Man may be the earliest known evidence for such ceremonies.[37]

Mungo Lady and Mungo Man were buried within sandy sediments forming the large lunette (dune) located along the eastern margins of Lake Mungo. Although the lake hasn't seen permanent water for nearly 15,000 years, when Mungo Lady and Mungo Man lived there it was filled with water and was a fertile environment very different from the semi-arid conditions of today. Both lived at a time of lowered sea levels, when you could walk from Victoria to Tasmania and when Sydney Harbour was a large forested inland valley.

Although Mungo Lady and Mungo Man were located only 450 metres apart, they were treated to two different mortuary traditions – one a cremation and burial, the other an extended body burial. It is impossible to know if they knew each other when they were alive: unfortunately, the resolution of dating is too imprecise to allow such a conclusion. They may have known each other, or they may have been separated by more than fifty generations. Whatever the case, it is clear that peoples living on the shores of Lake Mungo 40,000 years ago practised a range of mortuary practices, from cremation to full-body burials. That the two Mungo burials are not aberrant is revealed by the subsequent recording of over sixty other burials in the Willandra Lakes Region, many dating to the Ice Age.

Remarkably, these burials include adults and children; double burials, flexed burials, crouched burials (in a sitting position) and cremations; and ritual fragmentation of bones prior to interment.[38] Webb has identified the remains of seventeen people from Ice Age sediments in the Willandra Lakes Region that show signs of cremation.[39]

Webb rightly noted that the Willandra Lakes burials demonstrate that the lives of Ice Age residents of the region included 'ceremony and ritual'.[40] That these people had rich ceremonial and ritual lives is not a surprise – all human societies have such dimensions. What is remarkable is that archaeological evidence for such richness of ceremony and ritual has survived the passage of at least 40,000 years. The ceremony and ritual associated with Mungo Man is something with which we can all empathise. Archaeologist Scott Cane sums up this empathy eloquently:

> this is the first evidence in the world for emotive cognition and recognition of the metaphysical itself. The burial thus speaks to us in human and spiritual terms: a loved one dead, his community grieving, a society responding with the acquisition of materials required for proper burial, his body adorned and respectfully buried, his ritual status recognised and his transfiguration anticipated through adornment. The sense of ritual attendance, the decorative consideration and careful placement of the corpse implies grief, communal concern and reverence. The importance of the burial is not simply that it is the first of its kind in the world, but that it presents us with recognisable human emotions – and so creates an empathic

link between us and this ancient community across more than 40,000 years of time.[41]

The Mungo Lady and Mungo Man burials show that living relatives cared greatly for both people and invested considerable time and energy in their funerals and interments, and also in curating and caring for their graves over many generations. The fact that both were buried within the same lunette where people camped shows a special and indeed sacred relationship between the living and the dead and the importance of ancestral presence and ancestral connections to place. This is much more than a belief in an afterlife: it is about the living and the dead living together and caring for each other. These sentiments are familiar to all First Nations Australians today and are timeless. They also help outsiders better understand why local Traditional Owners have repatriated Mungo Lady and Mungo Man for reburial at Lake Mungo.[42]

The Mungo burials, along with the plethora of Ice Age camp sites with stone tools, fireplaces, and mussel shells and fish bones (e.g. golden perch) from ancient meals, touched the soul of Australia and placed our ancient past at an unprecedented level on the world stage. Such international standing was recognised in 1981 with the inscription of Lake Mungo and the broader Willandra Lakes Region on UNESCO's World Heritage List.

The internationalisation of Mungo Lady and Mungo Man and their connections with peoples in distant lands does resonate with life at Lake Mungo 40,000 years ago. The red ochre that was ritually sprinkled across Mungo Man is not of local origin, with the nearest potential source 200 kilometres distant.[43] In the next chapter we explore how the world of Ice Age Australians extended well beyond the local, with trade networks connecting communities located hundreds of kilometres apart.

4

ENHANCED NETWORKING, TRADING AND SHARING

Australian Indigenous societies innovated some of the world's most complex and extensive pre-industrial trading networks. Such innovation relates not only to the geographical scale and volume of materials traded, but also to the social and cosmological underpinnings of trade networks and provisioning expeditions, which in some cases are akin to religious pilgrimages. That is, religious, ceremonial and social connections among communities dotted along trade routes were often mirrored and celebrated in intricate Songlines enshrining elaborate Dreaming (cosmological) narratives of the creative actions of Ancestral Beings.[1]

It is for these reasons that Aboriginal trading systems extended well beyond the economic and the movement of utilitarian objects and materials from the haves to the have-nots. Aboriginal trade networks rarely involved movement of food items, unless it was for immediate consumption.[2] They often involved symbolic and even sacred objects that said more about cementing and celebrating religiously ordained (Ancestral Law) relationships among communities than about acts of survival. Indeed, the gift-giving dimension of Indigenous trade extended to procuring exotic objects that were locally available and even trading identical objects like for like.

Torres Strait Islanders developed an elaborate canoe-based trading system across their archipelago that extended north to include New Guinea Melanesians and south to engage with Cape York Peninsula Aboriginal peoples.[3] Anthropologist David Lawrence makes the important point that the Torres Strait trading system was not built around set trade routes but was more flexible and based on social relationships.[4] As such, trading pathways could change frequently depending on local circumstances. For example, trade relationships ceased for a period of time between islands when relationships broke down following a raid, especially if the raid included headhunting.[5] In addition, trade in food items (between islands and with the New Guinea mainland) was of greater importance in Torres Strait than on the Australian mainland.[6]

For mainland Aboriginal peoples, trade objects and ideas were literally walked thousands of kilometres from one side of the continent to another through hundreds of clan territories and from person to person. In 1950, anthropologist Johannes Falkenberg recorded in the Port Keats region of the Northern Territory one set of trade objects formally passing through the hands of

134 individuals over a distance of 200 kilometres.[7] In the early 20th century, ethnologist Daisy Bates spoke to numerous Aboriginal people across Western Australia who informed her about what she termed 'the great highway' associated with 'an ancient Aboriginal trade route [that] ran around the whole continent'.[8] This 'highway' or, indeed, these highways didn't appear out of nowhere. All have an ancient history and all express innovations in social networking and subsequent enhancements of social relations for strategic economic, political and ceremonial reasons. This chapter looks at innovations and associated enhancements in trade networks for two major time periods: the Ice Age (12,000–65,000 years ago) and the Late Holocene (the past 4000 years).

ICE AGE

What was the Indigenous population of Australia during the Ice Age? The short answer is that nobody knows, but we can make a few informed generalisations. As we discussed in Chapter 2, archaeologists have suggested that a founding population of at least 1000 people was required to people the continent of Australia.[9] Over the past 65,000 years, the number would have fluctuated. Ice Age Australia varied considerably through time in terms of environmental changes and land, water and food availability. Each of these factors affects population levels.

Two key environmental changes were linked to the peak cold and dry period of the Ice Age known as the Last Glacial Maximum (LGM), around 20,000 years ago. First, expansion of the arid zone during the LGM would have been conducive to population decrease, as the carrying capacity of many areas was far less than it is today. Ethnographic evidence shows that the population density of Aboriginal people in Central Australia (measured in individuals per square kilometre) was almost 100 times lower than that of coastal areas in northern and eastern Australia, indicating lower food production of arid regions.[10] Second, expansion of the Australian continent by 25 per cent due to lowered sea levels and exposure of the continental shelf during the LGM would have been conducive to population increase, as these vast coastal plains, especially in the more fertile parts of Australia, would have been capable of supporting large populations of people.[11] Although complex modelling indicates that historically known population levels may have been achieved within 6000 years of initial settlement, quantities of archaeological material suggest that such levels were not reached until the past 1000 years.[12] We suggest that the population of Australia during the Ice Age was much less than that found just before colonial invasion.

These figures are meaningful in the context of discussing Ice Age trade networks because long-distance movement of objects and ideas most likely involved fewer people and fewer trading partners located further apart than recent times. So what archaeological evidence exists to show that Ice Age communities located hundreds and even thousands of kilometres apart in Australia invested in being connected?

In terms of shared ideas, the best available evidence comes from rock art. For example, rock-art specialist and archaeologist Jo McDonald has identified remarkable widespread similarity in the creation of what are referred to as 'archaic faces' petroglyphs that were engraved into rocks across the north-west quadrant of Australia, taking in the northern half of Western Australia and western Northern Territory.[13] These rare and distinctive-looking faces with their piercing concentric-circle eyes are heavily weathered and are considered to have been made during the Ice Age over 20,000 years ago.[14] How is it that petroglyphs located up to 2000 kilometres apart are nearly identical in form and style? McDonald rightly notes that the artists who created these faces, which are spread thousands of kilometres apart across nearly 1 million square kilometres of what is largely arid country, were all part of huge 'long distance arid zone information exchange networks'.[15]

But what do traded objects tell us about the strategies developed during the Ice Age to support connection? Here we look at two informative object types: ochre, and shell body adornments.

Ochre

Ochre is found in the oldest layers of the Madjedbebe rock-shelter site, dating to 50,000–65,000 years ago, and has been recovered archaeologically from numerous Ice Age sites.[16] Ethnographically, it is known that Aboriginal and Torres Strait Islander ochre use had deeply symbolic and usually sacred associations.[17] Few ceremonies took place without ochre body adornment. Such symbolism is universal across the globe, and archaeologists use ochre as a marker of symbolic thought and ritual practices by early humans 100,000–500,000 years ago, including Neanderthals.[18] The oldest and most poignant example of ochre use in Ice Age Australia is associated with the 40,000-year-old Mungo Man burial in western New South Wales (see Chapter 3); however, the nearest known potential source of ochre is located 200 kilometres away.[19] Whether or not the ochre was obtained directly by Lake Mungo people or arrived through trade is unknown.

The best-documented example of ochre mining by Aboriginal people during the Ice Age is the Karrku ochre mine on Warlpiri County in the south-west of the Northern Territory. The site was used by men and women and continues as an active mine; its ochre is traded widely across the region.[20] The site has strong mythological associations and 'lies on the intersection of several Dreaming tracks'; one Dreaming narrative associates the ochre with the blood of an Ancestral 'slaughtered man'.[21] Ochre from Karrku is a deep red colour and has flecks of mica that give it a 'silvery sheen'.[22] The narrow entrance tunnel opens into a sizeable inner chamber measuring 12 × 9 × 1 metres, from which runs a 20-metre-long horizontal shaft and another smaller chamber. Anthropologist Nic Peterson and archaeologist Ron Lampert estimated that 240 cubic metres, equating to 300 tonnes, of ochre have been mined from the site.[23]

Numerous fragments of Karrku ochre used between 15,000 and 35,000 years ago were excavated by archaeologist Mike Smith from Puritjarra rock-shelter, 125 kilometres south of the ochre mine.[24] As such, the Karrku ochre mine began operations at least 35,000 years ago. Smith argues that the Ice Age occupants of Puritjarra obtained Karrku ochre directly themselves during their regular highly mobile lifestyle, and not indirectly through trade.[25] In contrast, archaeologist Kate Morse suggests that ochre fragments in layers dating to between 24,000 and 30,000 years ago at Mandu Mandu Creek rock-shelter on the central-west coast of Western Australia were likely obtained through 'long distance trading networks' from the nearest potential ochre source in the Hamersley Plateau, 300 kilometres to the north.[26]

Shell beads

Archaeological research reveals that Ice Age Aboriginal Australians developed some of the earliest and most elaborate examples of shell jewellery and body adornments in the world. For example, Kate Morse excavated twenty-two cone shells 'deliberately modified as beads' that formed a necklace from the bottom levels of Mandu Mandu rock-shelter dating to at least 36,000 years ago.[27] Although located adjacent to the sea today, the cone shells must have been obtained from the ancient shoreline that was 10 kilometres to the west during lower sea levels. Such proximity implies direct collection by the necklace's owners and not procurement through trade.

In the north of Western Australia, archaeologist Jane Balme excavated ten marine scaphopod (tusk) shell tubular beads dating to 31,000–33,000 years ago from Riwi rock-shelter. Similar tusk-shell beads are made by Bardi women of the Kimberley coast to the north-west. Today Riwi is located 300 kilometres inland, but during lower sea levels around 30,000 years ago, it was at least 500 kilometres from the sea.[28] Balme and Morse suggest that people camping at Riwi at least 30,000 years ago probably obtained the marine-shell beads through trade. They add that 'the existence of exchange networks' would have been 'important for maintaining relations between groups'.[29] Archaeologist Sue O'Connor similarly argues that a fragment of pearl shell excavated from Widgingarri Shelter 1 in the Kimberley and dating to 23,000 years ago was obtained indirectly through trade from peoples on the nearest Ice Age coast, which at the time was located 'over 200 km' away.[30]

Archaeological evidence for Ice Age Australia presents a similar picture of large-scale networking on a scale of hundreds of kilometres in terms of the movement of objects (e.g. ochre and marine shells), and thousands of kilometres in terms of the movement of ideas (e.g. 'archaic faces' petroglyphs). Much of this evidence supports in part archaeologist Harry Lourandos's view that Ice Age communities of Australia were 'more mobile' and part of 'more open social networks' than communities of the last few thousand years.[31] Indeed, much archaeological evidence for long-distance trade during the Ice Age in Australia is equivocal given that so-called exotic objects may have been obtained directly by their users, who were highly mobile with large territories. In the next section we will show how these cultural patterns differed fundamentally from large-scale procurement expeditions and transcontinental trade networks that developed over the past 4000 years during the Late Holocene.

LATE HOLOCENE

The Late Holocene – that is, the past 4000 years – witnessed dramatic enhancements in social networking and trade in objects and ideas across the continent. A rise in both the quantity of cultural items unearthed from sites (particularly in caves and rock-shelters) and the number of sites is indicative of a dramatic increase in the Indigenous population at this time in Australia.[32] This population increase was probably associated with growing numbers of clans and tribal groups. More groups means more social boundaries and an increased need to invest in innovative social and political strategies to help ensure everyone gets on.[33] A key strategy in this regard is enhancement of formal and informal social networks expressed through intergroup ceremonial gatherings and intergroup trade.[34] Another important element of the need for greater networking is the increased need for food security, especially if new groups take up less productive territories or become more sedentary such that periodic food shortages from drought, for example, cannot simply be dealt with by temporarily moving to neighbouring greener pastures. Now those greener pastures might be in the hands of a new group. However, by investing in new social networks (e.g. through new trading systems, marriage alliances and shared ceremonies), intergroup connections and friendships can be made and relied upon as 'safety nets' in times of need.[35]

In marked contrast to Ice Age trading systems, whereby objects moved across the landscape from person to person over hundreds of kilometres and ideas could be conveyed thousands of kilometres, the Late Holocene sees the innovation and development of more formalised contexts for trade, whereby people with objects and ideas to trade converged on well-known gathering centres. Many of these gatherings involved hundreds or even over 1000 participants, so detailed logistical planning went into selecting locations that could accommodate that many mouths to feed over what could be weeks or even months.[36] Some gathering centres became famous and were timed to take advantage of seasonal superabundances of certain foods, such as bunya nuts (south-east Queensland), eels and other freshwater fish (south-west Victoria, south-west Western Australia, north-central New South Wales), yams and cycads (northern Australia), grass seeds (Central Australia) and bogong moths (South Eastern Highlands).[37] In many cases, food superabundances were orchestrated, such as eel aquaculture and stockpiling of fish and grass seeds. These gatherings were multipurpose in that trade helped cement social, political and ceremonial alliances.

Late Holocene social connections were sometimes expressed less in the movement and trade of objects and more in the movement and sharing of ideas related to technology, rituals, ceremonies and Dreaming creation narratives, and Songlines.[38] Tula adzes are an excellent example of the movement and shared uptake of a distinctive and highly innovative stone-tool type across Australia's vast arid zone.[39] These uniquely shaped adzing tools were designed to work hardwood and continued to be used by numerous desert peoples into the 20th century. They are made by taking a thick, circular-shaped flake and neatly chipping its margins to create a chisel to allow careful and precise shaving and shaping of wood. All were hafted onto wooden handles with plant adhesive, mostly made from spinifex-grass resin.

No tulas have been identified by archaeologists in sites older than 4000 years, but after 4000 years ago these innovative tools start turning up in their thousands across the arid zone. They even start appearing in the Kimberley region on the north-west coast and on the central-east coast.[40] Since there are many ways to carve and shape wood, large-scale uptake of tulas reveals a desire of desert peoples to all be on the same page and to share in the use of this new and distinctive tool type. That is, tulas became fashionable and represented a desire of hundreds of communities for social inclusiveness and the use of a mutually recognisable object. Tulas were often made from local stone sources, but many, along with hafting resin, were traded between groups.[41]

The fact that Aboriginal communities of the central east coast also took up tulas reveals that the desire for large-scale social inclusiveness went well beyond the arid zone. Ian has suggested that the desire for social inclusiveness extended to other stone-tool types, such as microliths.[42] Microliths are small crescent-shaped to triangular-shaped tools made from flakes and local stone sources. Use-wear and residue analyses reveal that they were used for a wide range of cutting and scraping tasks.[43] Archaeologist Peter Hiscock notes that microliths 'proliferated' across the southern three-quarters of the continent (excluding Tasmania), taking in a wide range of environments, between 1000 and 3000 years ago.[44] Although some microliths were used 9000 years ago and perhaps at the end of the Ice Age, something dramatically changed such that use of these tools literally exploded across mainland Australia; use dropped off equally as dramatically in most areas over the past millennium.

Similar innovations in cultural sharing and social inclusiveness occurred over the past 3000 years in tropical north-east Australia, involving Aboriginal peoples of Cape York Peninsula, and Torres Strait Islanders and their northern neighbours across southern-central New Guinea. Ethnographic museums across the world contain a wide variety of objects taking in these three major cultural regions, with some objects found across all three (e.g. bamboo smoking pipes, double-outrigger canoes), some found only across New Guinea and Torres Strait (e.g. dog-tooth adornments, stone-headed clubs), and some only across Cape York Peninsula and Torres Strait (e.g. shell-handled spearthrowers, nautilus-shell headbands).

These objects were often traded between groups, but in most cases it was the idea of the object that moved, meaning that local communities made their own versions. Recent identification of the manufacture of Melanesian-style earthenware pottery by Aboriginal peoples on Jiigurru (Lizard Island) off the North Queensland coast reveals that such large-scale sharing of ideas across the Coral Sea dates to 2000–3000 years ago.[45] Ian has conceptualised this extraordinary tropical melting pot of shared ideas and complex networking as the Coral Sea Cultural Interaction Sphere.[46]

The Coral Sea Cultural Interaction Sphere is made up of a series of connecting maritime trading routes spread along 2000 kilometres of coastline. Historical records (archival and oral) and anthropological research indicate that terrestrial trading routes similarly crisscrossed the Australian continent. In 1939, ethnologist and archaeologist Fred McCarthy from the Australian Museum in Sydney published a remarkable paper synthesising historical and anthropological information on Australian Indigenous trade in objects and ceremonies.[47] He found that trading routes operated at differing scales, from myriad local routes among neighbouring groups through to what he called 'trunk trade routes' comprising multiple local routes that collectively covered 'astounding distances'. In this sense, the Coral Sea Cultural Interaction Sphere is a trunk trade route. Below we provide four examples of objects that moved great distances along inland trunk trade routes.

Ochre

Over the past 65,000 years, Aboriginal communities have used ochre from local and exotic sources. The Late Holocene witnessed an extraordinary ramping-up of use of exotic ochres procured through enhanced trade networks and the innovation of dedicated, large-scale mining expeditions. Ethnographic information is available for a number of remarkable Aboriginal ochre mines and associated long-distance expeditions.

The Thuwarri Thaa (aka Wilgie Mia) ochre mine in Wajarri Yamaji Country in the central west of Western Australia is spectacular.[48] It is the largest Aboriginal ochre mine in Australia and remains in use for body paint for ceremonies and to paint wooden artefacts. The red ochre, the primary colour at the site, is the blood of Marlu, an Ancestral red kangaroo, with yellow ochre his liver and green pigment his gall; the Ancestral spiritual power and potency of Marlu stays with the ochre after it is mined. It was originally mined using wooden scaffolds, fire-hardened digging sticks, and stone tools. Cavernous excavation includes a 30-metre-long subterranean gallery sloping down from the surface, and smaller galleries and crawl spaces that are 'up to 60 m long'.[49] In the early 1900s, Henry Woodward, assistant government geologist of Western Australia, calculated that 15,000 cubic yards – equivalent to nearly 11,500 cubic metres and 24,000 tonnes – had been quarried at the site.[50] Archaeological excavations revealed over 6 metres of mining spoil radiocarbon-dating to at least 1000 years ago.[51] Archival and contemporary records indicate that the ochre was carried 300–600 kilometres to numerous Aboriginal communities; today it is driven by car to Central Australian communities located up to 900–1100 kilometres from the mine.

Equally extraordinary are expeditions associated with obtaining permission from clan owners to mine ochre from the famous and sacred Pukardu (aka Bookartoo) ochre mine near Parachilna in the Flinders Ranges of South Australia.[52] This mine was intimately connected with a number of sacred Dreaming creation narratives, with the ochre variously associated with emu and dingo blood.[53] In the 1860s, Alfred Howitt recorded that in July or August each year, 70–80 Diyari men of the Cooper River region east of Lake Eyre made the 'perilous journey' up to 500 kilometres south to Pukardu for ochre. The quarried ochre was 'kneaded into large cakes' weighing 30–35 kilograms each and carried back home.[54] Local policeman Samuel Gason similarly recorded ochre expeditions by Cooper Creek Aboriginal men, adding that the six- to eight-week expeditions involved travelling 'about twenty miles [32 kilometres] a day' and that 'each man carries an average weight of 70 lbs. [30 kilograms] of ochre, invariably on his head'.[55] Each carrier stood on his head to press a shallow depression into his cake such that it rested more comfortably on his head with grass padding for the long walk back home.[56] These numbers equate to 2–3 tonnes of ochre per expedition. In addition to visitors from the greater Lake Eyre region 400–500 kilometres to the north, mining expeditions were organised by groups located over 1000 kilometres to the north-east (headwaters of the Diamantina and Georgina rivers and Charleville in Queensland), 500 kilometres to the north-west (Oodnadatta region in South Australia), 400 kilometres to the south (Adelaide Plains) and 500 kilometres to the south-west (Eyre Peninsula).[57] In many cases, once the expedition groups returned home they allocated some of the Pukardu ochre for trade on to neighbouring groups.[58]

Museum anthropologist Philip Jones points out that while the Pukardu expeditions were part of trade networks, with visitors bringing objects in exchange for the ochre, at a more profound level they were 'archetypal religious pilgrimages' and a 'rite of passage' for young male participants.[59] Anthropologist AP Elkin pointed out in the 1930s that the Pukardu expeditions followed the path of Dreaming creation narratives (Songlines) and were in a sense a 're-enactment of the past'.[60] Although archaeological information is unavailable on the long-term history of mining at Pukardu, archaeologist Mike Smith posits that large-scale mining dates to the past 500 years.[61]

Baler shells

The movement of oval-shaped objects made from baler (*Melo* spp.) shell, a tropical marine shell found along the coastline of northern Australia, into southern Australia provides an extraordinary example of the geographical scale of Aboriginal trade networks. Historical (archival and oral) records, along with ethnographic recordings, reveal that baler-shell pendants manufactured by a select group of coastal communities in North Queensland (south-eastern Gulf of Carpentaria and western Cape York Peninsula, and possibly eastern Cape York Peninsula) and north-west Western Australia (Port Hedland) were traded inland to Central Australia.[62] Gulf of Carpentaria baler-shell objects were traded south following the rivers of western Queensland to Lake Eyre in South Australia and even down to the Flinders Ranges, a transcontinental journey of 1500 kilometres.[63] The ceremonial and ritual use of baler-shell objects increased with traded distance from the coast. Torres Strait Islander–manufactured baler-shell pubic covers were traded northwards into south-west Papua New Guinea.[64] Archaeological examples of baler-shell objects traded more than 400 kilometres inland have been recovered from the Great Sandy Desert of Western Australia and radiocarbon-dated to 2200 years ago.[65]

Grindstones

Nearly all archaeological evidence on the manufacture and use of large seed-grinding stones, as used by arid-zone women, points to proliferation over the past 3000 years. The best-documented quarry sources of these grindstones is sandstone outcrops in the Lake Eyre–Lake Frome region of north-east South Australia.[66] The scale of quarrying operations for production of trade grindstones was massive. For example, archaeologists Isabel McBryde and Mike Smith estimate that between 700,000 and 1.1 million grindstones were manufactured from the nearly 20,000 cubic metres of sandstone removed from more than 1000 mining pits stretching for 1 kilometre along the Wadla wadlyu quarry located south of Lake

Frome.[67] Large-scale production of grindstones was for local use but mainly for entry into inter-regional trading networks that extended up to 500 kilometres north towards Mount Isa in north-west Queensland.[68] In the Western Desert during the Late Holocene, seed-grinding stones occur mostly around permanent water sources that hosted large inter-regional gatherings of peoples for a broad range of social, ceremonial and political activities.[69]

Stone axes

Although the world's oldest ground-edge axes occur in Arnhem Land and were part of an Ice Age axe tradition across far northern tropical Australia (see Chapter 2), remarkably, stone axes occur across southern Australia only within the past 4000 years and are unknown in Tasmania and most of the south-west of the continent.[70]

The Late Holocene proliferation of stone-axe use was accompanied by development of major axe trading networks linked to establishment of huge stone-axe quarries. For example, the Lake Moondarra metabasalt axe quarry near Mount Isa in north-west Queensland extends over an area of 2.4 square kilometres.[71] It features hundreds of stone extraction pits where up to 30 cubic metres of rock rubble was removed, and manufacturing zones with metre-high piles of flaking debris where blocks of stone weighing up to 100 kilograms were broken apart and shaped into axe blanks ready for export. Archaeologist Peter Hiscock estimates that as many as 800,000 partly made axes remain at the quarry site, which may have operated for at least 1000 years. Mike Smith suggests that over 1 million axes were exported from the site.[72] The local Kalkadoon people took the axes to markets located immediately outside their territory, where exchange took place with their neighbours. From here the axes were traded southwards down western Queensland and as far as South Australia over a distance of 1000 kilometres.[73]

Australia's most famous Aboriginal greenstone axe quarry is located immediately north of Melbourne in Wurundjeri Country. It is known in the local Woiwurrung language as Wil-im-ee Moor-ring, but Europeans named the location Mount William.[74] Axes from the quarry were highly prized and were exchanged for items including possum skins and other equally valued objects. Additionally, the trade served crucial social purposes, reinforcing familial ties and connections across Country and helping to bring groups together. Several Elders, including Billibellary, head of the Wurundjeri Willam clan, inherited the distribution rights and controlled permission to access the quarry. The Mount William quarry was still in use in the 1830s and 1840s, when Melbourne was established.[75] It was well-known historically, and numerous European observers wrote about it. In 1854 the Polish naturalist William Blandowski noted he had been told about the Mount William axe quarry by 'tribes 400 miles [over 600 kilometres] distant'.[76] William Barak, renowned Wurundjeri Elder and spokesperson, noted that:

> When neighbouring tribes wanted stone for tomahawks they sent a messenger to Billibellary [quarry headman] to say they would take opossum rugs and other things if he would give them stone for them. Billibellary's father when he was alive split up the stones and give it away for presents such as rugs, weapons, ornaments,

> belts, necklaces – three pieces of stone were given for a possum rug. People sometimes give presents in advance to get stone bye and bye.[77]

In the 1970s, Isabel McBryde undertook a major archaeological study of the Mount William axe quarry, combined with geological sourcing of axes in museums and meticulous reading of historical documents, to determine the geographical scale and social dimensions of associated trade networks.[78] Axes made from Wil-im-ee Moor-ring greenstone were identified across the western half of Victoria (taking in tribal/language groups of the Kulin nation) and up to 700 kilometres to the west into South Australia and a similar distance north-west into New South Wales. Remarkably, few axes were traded eastwards into the Gippsland region of eastern Victoria and the territory of the five clans of the Gunaikurnai people, who speak different languages to the Kulin nation.[79] McBryde argued that the Gunaikurnai were not participants in the Mount William axe trading network due to their well-documented limited social connections with Kulin nation groups. As such, participation in the Mount William axe trade network was an expression of existing social relationships. Indeed, trade network participants often had near-identical stone outcrops for axe manufacture but chose to import Mount William axes. The social role of Mount William axes is further revealed by minimal evidence for their use in chopping wood by trade recipients, a finding consistent with the axes serving more as ceremonial and prestige status items.[80]

The Mount William axe trade is an excellent example of trade in objects more for their symbolic and prestige value and less for their utilitarian value. That is, trade in these objects was more akin to reciprocal gift-giving that was focused on cementing new or maintaining existing social relationships between individuals and communities. The social and symbolic value of Mount William axes is similar to Late Holocene ochre trade, where groups procuring exotic ochre bypassed the option of using their own local ochre outcrops. In each case, whether stone or ochre, the local sources lacked the social and symbolic value, prestige and, in many cases, sacredness of the exotic materials. Indeed, the social bonding role of trade was in some cases so emphasised that groups would import objects that they had originally made, such as Torres Strait Islanders importing stone club heads of Torres Strait origin from lowland New Guinea peoples, and trading like-for-like, such as groups in the Northern Territory who traded stone blades for 'identical' stone blades.[81] Through complex trade relationships developed over thousands of years, First Nations Australians innovatively networked an entire continent, with international links to New Guinea (via Torres Strait Islanders) and Indonesia (via Makassans – see Chapter 6), long before introduction of the telegraph.

5

ENHANCED PLANTS AND ANIMALS

First Nations' relationships with their world were poorly represented in Western scientific theories of 'hunter-gatherers' living at the mercy of the environment, or by more romanticised 'noble savage' views of living 'in harmony with nature'.[1] Such a passive relationship with 'the environment' is fundamentally at odds with First Nations' philosophies of their intimate and spiritual relationship with the world they are immersed in and their rights and responsibilities to take care of and nurture their environment. This curatorial relationship with the environment was also built around a religiously inspired philosophy – namely, if people look after the environment, then the environment will look after people. Such a mutual relationship, a co-dependency, was infused with morality and an understanding of the potency of the environment, which was energised by Ancestral powers and spiritual life forces. In this sense, the environment, and its plants and animals, thought about people with an expectation that they had responsibilities to 'do the right thing'. Such philosophical views are at odds with Western constructions of the environment as an external entity filled with resources for people to exploit, often to the point of exhaustion – or even extinction, in the case of some animals.

The past fifty years have seen a fundamental turnaround in the way Westerners have come to understand the scale and capacity of First Nations' modification and construction of the Australian environment. Arguably the most powerful insight is an understanding of what archaeologist Rhys Jones evocatively termed 'fire-stick farming', whereby over thousands of years of cultural burning practices, Aboriginal peoples transformed plant and animal communities across Australia.[2] Such understandings have been advanced profoundly by the subsequent research of Bill Gammage, Bruce Pascoe and Michael-Shawn Fletcher.[3] The capacity to transform and curate environments to enhance sustainability of plant and animal use was an expression of long-term ecological knowledge gained through thousands of years of intimate observation and empirically based scientific conceptions of ecological processes.[4]

Indigenous notions of ecological processes extended well beyond traditional Western scientific views on the blind mechanics of evolution. For Indigenous peoples, ecological processes are also expressions of the creative powers and will of spiritual entities; these creative powers are engaged with, promoted and maintained through ritual and ceremony. Anthropologist Peter Sutton and archaeologist Keryn Walshe refer to such engagement as 'spiritual propagation'.[5]

In this chapter we have selected a series of case studies that illustrate the depth, breadth and innovativeness of Aboriginal and Torres Strait Islander long-term ecological knowledge. Following a brief overview of knowledge and partitioning of the seasons and creation of calendars to enhance contemporary management of lands and seas, we move to plant enhancements and ingenious removal of deadly toxins from plant foods to make them edible. We then explore the relationship between people and marine animals in terms of dugong and turtle rituals and animals that knowingly assisted people to hunt and fish. The chapter ends with new and challenging insights into farming practices, focusing on aquaculture in freshwater (eels) and saltwater (oysters) contexts.

SEASONAL KNOWLEDGES AND CALENDARS

Over millennia, Aboriginal people developed intimate understandings of seasonal patterns to secure food (plant and animal), medicines and other resources. They knew when to select grasses that were pliant enough to be woven into baskets, when fish or eels were plentiful, and when turtle eggs were ready for harvest. Seasons also signalled the timing for ritual and cultural activities. In this sense, seasonal structuring of the year was an expression of environmental change seen through the lens of cultural events.[6]

Australia is a complex country with a range of climate zones that do not easily fit into the fixed four-season calendar developed for Europe. In the early 2000s, Lynette worked with the Bureau of Meteorology (BOM) on the possibilities of developing seasonal calendars that tied together Indigenous weather and ecological knowledges.[7] Indigenous seasonal calendars that rely on these knowledges, including seasonal availability of specific plant and animal resources, have been developed by a range of Aboriginal and Torres Strait Islander communities and presented graphically on posters over the past twenty years.

Both BOM and the Commonwealth Scientific and Industrial Research Organisation (CSIRO) have been proactively engaging with Indigenous communities as they create novel ways of depicting the seasons specific to their Country. Indigenous seasonal knowledges and calendars also provide an innovative way to integrate traditional ecological knowledges into 'natural' resource management decisions.[8] The knowledges contained within these calendars have been used to augment Western scientific approaches to land and sea management. When combined with quantitative surveys of plants and animals (carried out by Indigenous rangers), the calendars' underlying ecological knowledge provides 'insights into harvesting and resource management strategies not revealed by the discrete time-bound surveys'.[9]

A visually impressive calendar was produced by the Wagiman community from Daly River in the Northern Territory. In a partnership that was sponsored by CSIRO, the Wagiman knowledge holders depicted the importance of waterbirds as markers of the wet season.[10] For the Wagiman, their early dry season makes hunting for goannas much more effective, and when the wet-season waters recede, the women know it is time to search and dig for hibernating freshwater turtles, which are then at their fattest. The depictions in the calendar include customary rules relating to animal collection and vegetation burning, all of which are done in a patchwork or mosaic fashion to ensure the people's use of Country is sustainable.[11]

Apart from an effective means of transmitting cultural knowledge to younger generations, Indigenous seasonal calendars allow Indigenous science to work with Western science to return a better balance to land and sea management, which ensures sustainable harvesting of resources, sustainable environments and sustainable communities.

FOOD-PLANT PRODUCTION AND FOOD TECHNOLOGIES

Like many people, during the COVID lockdowns we spent a great deal of time in our garden. While we have long had a small patch of vegetables and herbs, we had yet to venture into growing what the nurseries and seed sellers call native or bush foods. The pandemic and lockdowns offered us a chance to expand our vegie patch to include Indigenous food, so we duly ordered seeds for warrigal greens and murnong yam daisy, and planted a variety of finger limes and sweet shrub-berries known as muntries. We are two people who pride themselves on having green thumbs, but these bush foods certainly tested us. It seemed that despite our experience in gardening, our knowledge of native plants and bush foods was woefully inadequate. According to the seed supplier, warrigal greens (also known as Botany Bay spinach) are named for the seed, which resembles a dog's head – warrigal being the Wiradjuri word for dingo/dog. The growing instructions assured us that these were easy to strike and we could expect a large crop of nutritious Botany Bay spinach within a few months. We tried multiple ways of germinating the seeds: in garden beds, in hydroponic pots and planted in the ground. Nothing worked, and we had to declare the warrigal greens a failure. Murnong on the other hand produced a delicious, though small, crop of yams. At present our garden has muntries, and finger-lime plants that are yet to yield any fruit.

As Pascoe and Gammage illustrate in their book *Country: Future Fire, Future Farming*,[12] across Australia and over thousands of years Aboriginal people managed their landscape through fire and firestick farming. The expert use of fire ensured that preferred animals and plants were available in plenty. In *Plants: Past, Present and Future*,[13] co-authors Zena Cumpston, Michael-Shawn Fletcher and Lesley Head note that Indigenous knowledges of edible, medicinal and useful plants were fundamental to life: plants provided food, clothing, shelter, tools and medicine. This, too, was knowledge honed over generations and based on millennia of observation and experimentation.

Yams were a desired food across many parts of Australia. Ethnobotanist Beth Gott describes the murnong yam daisy as one of the main plant foods for Victorian Aboriginal people.[14] William Buckley, who lived for thirty-two years among the Wadawurrung people in the area now known as Geelong, described 'roots' as being a staple food.[15] Yams or roots are a reliable food source as they are available almost year-round. European explorer Thomas Mitchell in 1839 observed vast plains of the yellow flowers of the yam daisy in western Victoria.[16] The following year, in the same area, Chief Protector of Aborigines George Augustus Robinson described 'millions of murnong or yam all over the plain'.[17] Yams were also cultivated in South Australia and Western Australia, where they were also stored for future consumption.[18]

Detoxifying plants

Across Australia, Aboriginal people have a long history of using plants for medicinal purposes, including those that are toxic. Over generations they accumulated knowledge and developed methods for processing toxic plants to make them safe for consumption. The methods included cooking, fermenting, soaking, leaching and roasting. These techniques can help to reduce or eliminate a plant's toxicity and make it safe to eat. For example, cycad seeds are so toxic that even touching them can result in poisoning, but Aboriginal people in northern Australia processed the plant by leaching then roasting the seeds to remove the toxic compounds; after that, the seeds could be ground into a starchy flour and used in cooking.[19] Early European explorers observed that cycads were an extremely important food source, particularly in times when large gatherings of people for important ceremonies were held.[20]

The Yanyuwa people of the south-western Gulf of Carpentaria and their neighbours the Garrwa were specialists in the processing of poisonous cycad palms. Anthropologist John Bradley has shown that the processing of cycads for these communities was not only about providing food, but that the labour-intensive methods were also about social organisation, identity and place, connections and relationships.[21] Previous researchers have emphasised that cycads and other toxic plants that needed elaborate processing to make them edible would only have been consumed in times of food shortages, or as a 'communion food' when they were needed to feed large groups of people who had come together for ceremonies.[22] For the Yanyuwa and Garrwa, Bradley argues that the relationship to the cycad is more complex: 'The cycad food was once eaten, but it requires correct preparation before it can provide sustenance; conversely, if it is used incorrectly it causes illness and ultimately death. This perhaps is the metaphor for Indigenous perceptions of land use in general.'[23]

Understanding how to remove the toxins from cycad seeds is not merely a matter of making them edible: this knowledge is about living on and engaging with Country. Even though cycads are no longer eaten on the Yanyuwa and Garrwa lands, some groups still have rights and responsibilities to curate extant cycad groves. These include obligations in the form of ceremonial knowledge, and the importance of regularly burning Country. Those who have these rights must ensure cycad palms are not damaged, and even if the food is no longer consumed it still must not be wasted.[24]

Rainforest Aboriginal peoples of north-east Queensland processed a range of toxic plants for food, including cycads. According to archaeologists, rainforest clans have been using poisonous plants for at least 1500 years and potentially as much as 2500 years.[25] The importance of plant foods to traditional precolonial diets should not be understated. Both written and oral histories show that plants accounted for a major part of Indigenous diets across Australia prior to European invasion. In north-east Queensland, several hazardous rainforest tree nuts were collected, processed and consumed with other rainforest plants and animals as part of traditional subsistence practices. The ethnohistorical literature often describes the careful preparation and eating of such nuts. This along with archaeological evidence has led to the speculation that toxic tree nuts were a dietary staple for Aboriginal rainforest people.[26] For example, various techniques were used to render cycad seeds safe to eat.[27] Once the safety of the seeds has been determined, they can be roasted or boiled and consumed as a starchy food – even baked into a 'bread'. Grindstones associated with cycads are commonly located in rainforest archaeological sites.

MARINE RITUALS

Deep connections between coastal communities and marine animals have manifested in marine rituals for dugongs and turtles, and in cooperative practices with animals that assisted people to hunt and fish. In many regions, a thriving Indigenous aquaculture focused on farming in both freshwater and saltwater settings. Here, we explore the elements of these enduring relationships, rituals and practices.

Dugongs

Numerous coastal communities across the tropical Indo-Pacific region hunt dugongs for food.[28] Torres Strait Islanders and a number of Aboriginal coastal groups across the northern half of Australia are experts on dugongs and have used that scientific knowledge for thousands of years to ensure sustainable hunting of these marine mammals. Archaeological research with Torres Strait Islanders has revealed dugong bones in old settlements that demonstrate hunting of dugongs to at least 4000 years ago.[29] Legendary narratives of Torres Strait Islanders from the Western Islands attribute the invention of dugong hunting – including introduction of hunting platforms erected over seagrass beds and use of the harpoon – to the Ancestral culture hero Sesserae.[30] For Torres Strait Islanders, knowledge of dugongs and the skills to successfully find and hunt these animals included spiritual and ritual domains. Indeed, Torres Strait Islanders developed the most elaborate dugong-hunting rituals documented in the world.

Torres Strait Islander hunters have a deeply intimate spiritual relationship with dugongs. The animals are also an important totem for many people in Torres Strait. Hunting of dugongs is associated with a wide range of ritual preparations and taboos that established communication between hunters and prey to facilitate a successful hunt.[31] This communication also involved a dialogue between the living and dead as it included connecting with the spirits of dugong hunters of the past (by talking to divination skulls in the past and visiting contemporary cemeteries) and spirits of dead dugongs (through respectful treatment of the bones of previously hunted dugongs).[32] In terms of the latter, little anthropological information is available and most published insights have come from archaeological research by Ian working closely and collaboratively with the Goemulgal (Western Islands) and the Kulkalgal (Central Islands).

Archaeological insights into dugong-hunting rituals have come from investigations of mounds comprising large collections of the bones (mostly skulls and ribs) of hunted dugongs. The number of dugongs represented in these mounds range from a couple of hundred to thousands in the case of the bone mound at the ancestral village site of Dhabangay on the island of Mabuyag.[33] Radiocarbon dating of the base of mounds suggests that most were created over the past 400 years. A remarkable aspect of these bone mounds is the number of dugong ear bones that have been deliberately extracted from skulls and placed, sometimes in pairs, within different levels of the mounds.[34] Information on dugong-hunting rituals is mostly kept secret, but what is public is that the ear bones were used by hunters to communicate with dugongs.[35] Another form of communication was more indirect and involved hunters letting dugongs know, through the careful placement of bones of previously hunted dugongs in bone mounds, that they are moral hunters with integrity. In this sense, hunters are letting dugongs know that they are treating their dead – the dugongs' ancestors – with respect.[36]

Turtles

On his first trip to Torres Strait in 1996, Ian was invited to meet up with the Goemulgal community of Mabuyag. A walk through the settlement of Bau takes you past St Mary's Church, from which emanate beautiful choral voices of women singing every Sunday morning. At the foot of the stairs leading into the church is a curious spherical boulder of granite that would require at least two strong people to move. Inquiries revealed that this boulder is the sacred wiwai stone from the turtle-hunting shrine at the ancestral village site of Goemu, located 1.7 kilometres to the south-west. Cambridge anthropologist Alfred Haddon visited Mabuyag in 1888 and 1898 and photographed the wiwai shrine with the spherical stone surrounded by a number of small granite columns.[37]

Torres Strait Islanders and their Aboriginal neighbours along the east coast of Cape York Peninsula undertook a wide range of rituals associated with increasing success in turtle hunting. These Aboriginal communities also performed rituals to increase the availability of turtles, as did other coastal Aboriginal peoples of northern Australia. For example, John Bradley has documented numerous turtle ritual maintenance sites of the Yanyuwa across the islands of the Sir Edward Pellew Group. These sites are activated by ritual specialists brushing green foliage across certain rock outcrops. Morality and respect are important dimensions of maintaining turtle stocks, with hunters knowing that they must abide by strict cultural protocols when disposing of turtle remains as 'wrong disposal' will result in 'depletion in the number of turtles'.[38]

In the 1840s, artist Oswald Brierly, on board the HMS *Rattlesnake* during its hydrographic survey of Torres Strait on behalf of the British Admiralty, recorded that Kaurareg/Gudang peoples of south-western Torres Strait and the adjacent mainland coast around Cape York constructed 'piles of stones and turtle heads at different points where they look out for turtle. They say that these heads will bring the other turtle about. They call these places agoo [agu].'[39] Haddon was informed that on the Kulkalgal island of Poruma in the Central Islands is a zogo (ritual) shrine 'to ensure the abundance of turtle' comprising a giant clam shell containing a 'globular black stone, babat' that was activated during the surlal turtle-hunting season in November by the application of 'turtle oil'.[40] Turtle-hunting rituals were elaborated by Torres Strait Islanders to include a wide range of shrines and portable charms, the latter often attached to canoes before hunting expeditions.[41] Shrines often included special stones similar to the wiwai shrine on Mabuyag. As seen with dugong hunting, Haddon noted that 'success in catching turtle depended on the help of the spirits of deceased men, especially those of former successful hunters'.[42]

Archaeological research on turtle-hunting shrines commenced in 1984 when English archaeologists David Harris and Barbara Ghaleb-Kirby excavated the wiwai shrine on Mabuyag and noted that the spherical stone was missing and the columns were lying flat.[43] The age of the shrine was not determined. In contrast, archaeologist Sue McIntyre-Tamwoy excavated a turtle-hunting shrine located on Moebunum (Tree Islet) off the tip of Cape York in 2007 and radiocarbon-dated turtle bones to around 400 years ago. It is highly likely that much older turtle shrines exist in the broader Torres Strait region, given that excavations at Dhabangay ancestral village on Mabuyag by archaeologist Duncan Wright recovered turtle bone dating back 6000–7000 years.[44]

MARINE COOPERATION

For some Indigenous Australians, the social world of marine hunting and fishing included cooperation with other animals.

We illustrate the diversity of this cooperation with two examples: suckerfish or remora cooperation in turtle hunting, and dolphin/orca cooperation in fishing and whale hunting.

Suckerfish

Historical and ethnographic records from many parts of the world report the ingenious use of suckerfish or remora to assist with the hunting of marine turtles.[45] Remora are caught, tied to a string line, and released from a boat to seek out and attach themselves to nearby turtles to aid their capture by hunters. No archaeological evidence is available for use of remora to hunt turtles; the oldest known account is by Christopher Columbus off the coast of Cuba in 1494.[46]

Aboriginal and Torres Strait Islander communities of north-east Australia provide some of the most detailed accounts of turtle hunting with remora. An early European recording of the technique comes from John MacGillivray, naturalist on board the HMS *Rattlesnake* in the 1840s. MacGillivray was informed by Giom (aka Barbara Thompson, the sole survivor of a shipwreck, who had been taken in and cared for by the Kaurareg Aboriginal people of south-western Torres Strait), that:

> A live sucking fish … having previously been secured by a line passed round the tail, is thrown into the water in certain places known to be suitable for the purpose; the fish while swimming about makes fast by its sucker [located on the top of its head] to any turtle of this small kind which it may chance to encounter, and both are hauled in together! [into a canoe][47]

Coastal Aboriginal groups of the Tully River region of the North Queensland coast south of Cairns used remora to catch not only marine turtles but also fish and dugongs.[48] Caught remora were kept alive in water within a canoe or in a bark container. Prior to heading out to sea for hunting, twine was tied to the remora's tail such that its location would be known after it attached to a fish, dugong or turtle. The prey was then pulled towards the canoe to a point where a spear or harpoon could be dispatched. In the same region of the Great Barrier Reef, naturalist Edmund Banfield observed that Aboriginal people kept tethered remora attached to the bottom of their bark canoes and gave them a tug when ready to find a turtle.[49] Once a remora had attached to a turtle, 'telegrams along the line from the sucker give precise information' to hunters in the canoe on the turtle's movement.[50] Similarly, in Torres Strait turtle hunters know when gapu (remora) have attached to a shark and not a turtle by subtle differences in tension of the string. [51]

Badhulgal of Badu in central-western Torres Strait and the Kaurareg of Muralag in the south-western Strait informed Haddon that the Ancestral culture hero Bia invented the technique of using suckerfish to catch turtle. According to legendary narratives of the Meriam of Mer in eastern Torres Strait, Barat of the island of Mua (located immediately east of Badu) 'taught the Western islanders how to catch turtle with the sucker-fish'.[52] Haddon added that Bia was known as Barat when he came to Mer.[53]

Some turtle hunters told Haddon that gapu possessed 'ominous powers' and would inform hunters on the condition of their canoe (as seen from underwater) and indicate problem areas (e.g. bow, stern) based upon where the gapu attached itself to a turtle carapace.[54] In this sense, Torres Strait Islanders' relationship with suckerfish extended beyond simple exploitation to one of mutual respect, whereby hunters recognised gapu sentience and the creatures' desire to volunteer advice on canoe condition. Such recognition and respect indicates that Torres Strait Islanders saw suckerfish as active participants in their society.

Dolphins

Various coastal communities around the world fished with the aid of dolphins. In most cases, dolphins were called up by fishers (e.g. by slapping the water with spears) and willingly and knowingly herded schools of fish, usually mullet, towards the shore for capture with spears or nets.[55] For Australia, fishing with dolphins has been documented for the Butchulla of K'gari (Fraser Island) and the Bunjalung of northern New South Wales, and for Moreton Bay in south-east Queensland.[56] Nineteenth-century accounts of cooperative fishing with dolphins for the central east coast of Australia point out that Aboriginal fishers had deep respect and affection for their dolphin fishing assistants. Infamous castaway Eliza Fraser observed that the Butchulla of K'gari 'almost deify' the dolphins that drove 'the fish toward the beach', and that 'it would be death … to kill or injure one of them'.[57] In some cases, Aboriginal people would 'sing' (ritually call up) dolphins to assist with fishing. For example, archaeologist Sarah Martin was informed by Aboriginal people of the Eyre Peninsula of South Australia of 'a special fishing place' where 'they used to sing sharks and dolphins to drive fish in towards the people on the shoreline where they surrounded the fish and picked them up'.[58]

Whaling at Twofold Bay, near Eden on the south coast of New South Wales, was remarkable for the role played by orcas (also known as killer whales). Orcas are the largest species in the dolphin family. Migrating humpback and southern right whales would swim past Twofold Bay on their way to and from their summer feeding grounds in the subantarctic waters of the Southern Ocean, and were easily herded by hungry orcas into the bay, where they had no way of escaping. Surveyor and ethnographer Robert H Mathews in the late 19th century interviewed Yuin people whose descendants today maintain strong relationships with the region's orcas. Mathews explained that the Yuin had used orcas to assist them in harvesting large whales. An Elder, usually a man, would construct a fire on the shore that was intended to attract the curious orcas, and:

> He then walks along from one fire to another pretending to be lame … leaning on a stick in each hand. This is to excite the compassion of the killers and induce them to chase the whale towards that part of the shore in order to give the poor old man some food. He occasionally calls out [in native language] … Heigh Ho … that fish upon the shore throw ye to me. [Once the larger whale is stranded] … men who have been hidden … make their appearance and attack the animal with their weapons. A message is sent to all their friends and fellow tribesmen in the neighbourhood inviting them to attend and participate in the feast.[59]

When European whalers moved into the area, they, too, took advantage of the orcas' herding behaviour, having almost certainly witnessed the Yuin 'hunting for whales' in this way. Right up to 1930, Old Tom, a male killer whale and pack leader, would herd baleen whales into Twofold Bay where the local whalers would harpoon and harvest them.[60] Today, replicas of the last three orcas used in whale hunting – Old Tom, Humpy and Kinsche – are suspended from the ceiling of the Great Southern Land environmental history gallery at the National Museum of Australia.

AQUACULTURE

Eels

The most famous example of Indigenous aquaculture in Australia comes from the Budj Bim Cultural Landscape World Heritage Area of the Gunditjmara in south-west Victoria. European understandings and appreciation of the existence of south-west Victorian First Nations aquaculture and its geographical scale, complexity and antiquity slowly developed over the past 180 years. The first European insights came from Chief Protector Robinson in 1841 when he was travelling across low-lying ground east of Gariwerd (the Grampians). In his journal he recorded coming across:

> an immense piece of ground trenched and banked, resembling the work of civilised man but which on inspection I found to be the work of the Aboriginal natives, purposely constructed for catching eels. A specimen of art of the same extent I had not

> before seen and therefore required some time to inspect it, and which the absence of transport enabled me to do. These trenches are hundreds of yards in length. I measured at one place in one continuous trepple [triple] line for the distance of 500 yards [460 metres]. These treble watercourses led to other ramified and extensive trenches of a most tortuous form. An area of at least 15 acres [6 hectares] was thus tracd [sic] over.[61]

Here Aboriginal people had excavated thousands of tonnes of dirt to construct artificial drainage channels as extra habitat for eels, and probably other freshwater fish, and to aid containment and capture. Robinson's observations fell on deaf ears, as British colonists were not ready to ditch the concept of terra nullius and accept that they had invaded the lands of people who had cultivated the soil, let alone people who had engineered a landscape for aquaculture. It took another 130 years before Europeans finally accepted Aboriginal fish farming in south-west Victoria. This acceptance started with archaeological research led by Peter Coutts from the Victorian Archaeological Survey, who produced detailed maps of large stone-walled eel-trapping facilities in the Tae Rak (Lake Condah) district of Gunditjmara Country.

In addition to the trapping facilities, Coutts noticed that some channels fed floodwaters and eels into depressions that he called 'ponds' for storing eels.[62] At the same time, archaeologist Harry Lourandos began investigations of an extraordinary 3-kilometre-long excavated channel near Toolondo to the immediate west of Gariwerd. Lourandos showed that the channel artificially fed eels from one lake to another to create an entirely new location for them to grow and live.[63] A decade later, he felt confident to mention the 'F' word: the Toolondo and similar facilities 'could be viewed, in effect, as eel "farms", or at least managed eel habitats'.[64] Further detailed mapping of eeling facilities in the Tae Rak region, including holding 'pens', by archaeologist Heather Builth cemented the view that the Gunditjmara had engineered an entire volcanic landscape and associated wetlands for eel 'aquaculture'.[65]

Although this was not news to the Gunditjmara, what remained unknown was the antiquity of the aquaculture system across Budj Bim. To shed light on this question, the Gunditjmara teamed up with Monash University and recommended archaeological excavations at Muldoons Trap Complex on the south-east corner of Tae Rak. Under Ian's direction, excavations in 2006 at a basalt-block dam wall revealed that its foundations were created around 500 years ago. Follow-up excavations in 2008 at a nearby excavated channel revealed that it was filled with around 50 centimetres of flood sediments. Radiocarbon dating of charcoal fragments within the lower flood sediments revealed that the channel was created around 6600 years ago, making it the oldest known stone-walled fish trap in the world.[66]

The Gunditjmara point out that the excavated channel is part of an aquaculture facility, because it functioned as much more than a fish trap. That is, the channel was one component of a bigger system of channels and ponding areas that artificially controlled water flows, and hence eel movements and eel living conditions. In this sense, the 6600-year date can be associated with aquaculture and proves that Muldoons Trap Complex is the world's oldest known aquaculture facility.[67]

Keen to showcase their extraordinary culture and heritage to the world, the Gunditjmara embarked on getting the Budj Bim Cultural Landscape onto UNESCO's World Heritage List. After a long campaign led by Gunditjmara Elder Uncle Denis Rose and backed by the Victorian and federal governments, Budj Bim Cultural Landscape was World Heritage–nominated in 2018. After a year of detailed international scrutiny by UNESCO, it was successfully included on the World Heritage list in 2019. This was Australia's first World Heritage nomination to be led and developed by a First Nations community.[68]

Oysters

Oysters have been eaten by First Nations Australians for thousands of years. Tens of thousands of midden sites have been recorded along Australia's 34,000 kilometres of coastline, and many of these middens contain oyster shells from past meals.[69] The scale of past oyster use is exemplified by the Booral Shell Mound in Butchulla Country on the mainland coast opposite K'gari (Fraser Island) in south-east Queensland. Archaeological excavations of this mound led by Ian in 1989 revealed that it was used between 3100 and 800 years ago and contains shells from an estimated 6 million oysters.[70] No evidence exists for Indigenous overexploitation of oysters in Australia's past; however, the sustainability of oyster harvesting changed dramatically with the establishment of colonial oyster fisheries. In south-east Queensland, for example, annual harvesting peaked at 44 million oysters in 1891, 'followed by rapid decline and functional disappearance of the fishery' such that today the commercial harvest is only 2 million oysters per annum.[71]

Is it possible that high harvesting rates of oysters by Aboriginal people in the precolonial past remained sustainable due to enhancement of 'natural' oyster beds? For example, in Moreton Bay, south-east Queensland, Aboriginal people of North Stradbroke Island continue their ancestral tradition of 'oyster farming'.[72] Oyster aquaculture was built around encouraging oyster growth and fat content by moving small oysters into peak growing conditions in deeper water, restocking depleted beds, and extending habitats through construction of artificial reefs ('islands') on top of natural high points on the seabed using used oyster shells to catch spat. Gaps in the deposition of oysters in local midden sites suggest use of oyster shells to create artificial reefs going back at least 1000 years.[73]

It is entirely possible that the boom in harvests in the early years of the colonial oyster fishery was aided by Aboriginal aquaculture practices. Once Aboriginal people were unable to tend to their oyster 'farms' due to colonial invasion, stocks began to decline, as did the colonial oyster fishery. Although Aboriginal people such as the Butchulla were employed in colonial fisheries, their knowledge of oyster cultivation was largely ignored.[74] In the next chapter we explore how in other contexts First Nations Australians innovated strategies to operate within the colonial economy.

6

OUTSIDE VISITOR OPPORTUNITIES

When Europeans arrived in Australia, they were variously met with overt and covert resistance, retaliation and accommodation. In western Victoria and Tasmania, the early interactions are rightly described as battles, skirmishes and even war. Over the past century or more, historians have vigorously debated the topic of whether Australia was settled peacefully. The traditional and conservative narrative depicted British colonists and pioneers as settlers who established the colony without substantial violence or resistance. In recent decades, however, this story has been challenged and a great deal has been written about the ways in which Indigenous people resisted. Historian Henry Reynolds has, over many books, disputed the popular image of Australian colonisation as peaceful and demonstrated that there was widespread violence against Indigenous Australians.[1] In this chapter we consider three sets of early interactions that move beyond the question of whether there was resistance or accommodation as we look at innovative ways of dealing with newcomers.

DEALING WITH NEWCOMERS

In the following we consider a range of what we are calling 'economic innovations'. Some of these were imposed, some were enthusiastically adopted, and some developed out of long-term engagements with outsiders or newcomers. It is undeniable that for most Australian Aboriginal people the impact of colonialism was devastating – dispossession, dislocation, disease, murder and missionisation. However, there is another, largely untold story of Australian colonial history. This is a story of innovation, agency, enterprise and entrepreneurship. Numerous innovative economic strategies were adopted by Indigenous people in the wake of European colonisation. As access to their traditional lands and waterways was increasingly obstructed, some groups began to work alongside the colonists in an attempt to maintain their connections to Country. Across Australia, Aboriginal people (most often men) worked in the pastoral industries and agriculture; they were also present in mining operations, whaling and timber-getting.[2] Torres Strait Islanders were engaged in the pearling and fishing industries of northern Australia.[3] In this section we consider a range of innovative economic strategies, some of which are often overlooked or disregarded.

The groundbreaking work of noted historian Ann McGrath on Indigenous participation in the cattle industry has demonstrated that Aboriginal people – mostly, but by no means exclusively, men – were involved from the beginning of the colonial period. Their knowledge of Country, terrain, water supplies and animal behaviour was essential in setting up various cattle runs. However, their roles and impact were often downplayed and they were marginalised and exploited. Many were forced to work as cheap labour on pastoral stations, their knowledge and skills not adequately valued. Despite this, McGrath has demonstrated that some Aboriginal people were able to carve out a niche in the industry and establish themselves as skilled stockmen and stockwomen. By understanding the cattle industry and Indigenous participation within it, McGrath has emphasised the complex and sometimes contentious ways that Aboriginal people have adjusted to shifting social and economic circumstances.[4]

Other industries that Aboriginal people were attracted to include timber-cutting, farm labouring, work on the goldfields and, in the north, pearling and fishing.[5] Some were drawn to a life at sea. Many of these industries allowed Indigenous people to remain on Country, where they maintained their responsibilities and continued cultural practices. These are stories of Indigenous Australian peoples seizing the opportunity to profit from participation in the colonial economy. We do not wish to downplay the degree of coercion and even violence that accompanied these industries, but in many cases the people involved maintained and exercised a degree of personal autonomy and agency within their new changed circumstances. They acted and reacted, and sometimes they made unexpected decisions.

History has often labelled these people as victims, disempowered slaves or even indentured servants.[6] We prefer to see the creative and ingenious ways that their choices allowed them some freedom. The following are examples of economic innovation.

INNOVATIVE ECONOMIES

Tolls and begging

The arrival of Europeans in the late 18th and early 19th centuries introduced Aboriginal people to many new material-culture items. In Naarm, now known as Melbourne, Kulin people used what the Europeans called 'begging' as a means to secure money and goods. Begging was an innovative and justifiable tactic as the Kulin were in a sense requesting compensation for what the Europeans had taken from them. On the streets of Melbourne from the beginning of first settlement in 1835, Europeans disapprovingly observed and commented on Aboriginal people begging and 'soliciting sixpences'. This practice continued for decades. Even after the establishment of reserves and missions in the 1860s, begging was a useful strategy.[7] In the 1880s, members of the Board for the Protection of the Aborigines visited Coranderrk Aboriginal Station at Healesville, just north of Melbourne, and observed a remarkable continuity. According to a contemporary report in the *Age* newspaper: 'An ancient warrior [known as] … Pretty Boy … [His] principle [sic] acquaintance with the English language seemed to consist of being able to say "Gib it tickpen", and until that coin was handed over the visitors knew no peace.'[8] Given it lasted for decades, 'begging' was clearly an effective strategy for securing what was wanted.

Everyday contact with what the *Argus* newspaper characterised as 'begging' – which, as Richard Broome notes, was more likely seen by Aboriginal people as a reciprocal trade for their eviction and relocation, and disruption of their customary hunting practices – upset white settlers in Melbourne.[9] However, some who were much more engaged with the local communities, such as Chief Protector of Aborigines George Augustus Robinson, recognised that if Europeans were hunting kangaroos that were food for the Kulin, why then would the Kulin not be entitled to 'hunt sheep'?[10]

Like many other groups, the Kulin peoples of Naarm had practised gift-giving as a way to create alliances and obligations, and gift-giving or exchanges were a feature of early Victorian contact relations. Europeans were not unfamiliar with this practice; indeed, John Batman's illegitimate treaty was based on the settlers' assumption that they were entering into gift-giving in exchange for land.[11] Gifts were regularly utilised by the colonisers to establish ties, commitments and alliances with Kulin peoples. When NSW governor Richard Bourke visited the Port Phillip colony and the newly founded settlement of Melbourne in March 1837, he distributed blankets and clothes and gave out gifts of 'four brass plates as honorary distinctions for good conduct'.[12]

Early squatter EM Curr commented on Kulin agency and autonomy when he noted that there were men cutting wood and selling it for both money and resources. He also noted that Aboriginal women were employed as domestic servants. Curr recalled that in

1839 Melbourne Aboriginal people were a feature of street life. He wrote that:

> These once free-born lords of the soil seemed to make themselves useful under the new *régime* by chopping firewood, bringing brooms for barter, and occasional buckets of water from the Yarra; and might be seen a little before sundown retiring to their camps on the outskirts of the town, well supplied with bread and meat ...[13]

Mostly this employment tended to be seasonal and cyclical.[14] As with the pastoral industry, these engagements enabled Aboriginal people to remain on or near to their Country. Outside of Melbourne, Aboriginal people found work as trackers and native police, and there were various opportunities on the goldfields.[15]

According to Curr, both the Kulin and the settlers benefited from these economic arrangements. For the Aboriginal people these arrangements facilitated access to certain European goods, while the settlers obtained useful objects and materials. Assistant Protector of Aborigines William Thomas commented in September 1840 that Aboriginal people were securing all they needed in Melbourne and this made relocating them to distant sites much more difficult.[16] Thomas was attempting to move the Kulin people from the township to a station at Arthurs Seat on the Mornington Peninsula, 75 kilometres from Melbourne, but begging – the innovative strategy for acquiring food and provisions – was sufficiently attractive to many Kulin that persuading them to move proved difficult for Thomas.

William Adeney was a contemporary of Curr and Thomas who in 1842 had recently arrived from London and intended to 'take up' land in western Victoria. In his earliest days in Melbourne, he made a diary entry of a scene he witnessed:

> I was sitting writing a letter the other day and rose to peep through between the blind and window frame to see how the day looked out of doors when at the same moment a black … face suddenly came into very close proximity to mine but on the other side of the glass. It was that of an old native woman who … no doubt wished to see through the same aperture what was inside. As it happened I was … startled and could not imagine for a moment what it was. The old woman was as much surprised as I was and after gazing with open mouth a few seconds said boro boro but what she wanted I could not understand … [they are often seen] accosting passers by with 'give me black money' and various other similar expressions begging bread.[17]

The use of the term 'boro boro' for begging is telling. To borrow is generally used to mean taking and using something that belongs to someone else. Aboriginal people were clearly aware that the presence of Europeans and the development of the city of Melbourne had disadvantaged them; their lives had been fundamentally changed. While they had been dispossessed without payment, they nonetheless expected recompense – in fact, they demanded it.

In the first few decades of European settlement there were few options for paid employment, though some Aboriginal people found work in tanneries, as farm labourers, and as bullock drivers.[18] To get the Kulin to attend the first government station, missionary George Langhorne distributed food, clothing and blankets in 1836. He emphasised that they would earn more rations if they chose to put in a few extra hours of labour. Langhorne was sure that the Kulin would learn to value labour as a means of obtaining goods and that this process would integrate them into a European economic system,[19] but his attempts to impose economic engagement largely failed. When Aboriginal people themselves exerted autonomy and control, they were more likely to succeed.

Some of the Kulin showed a particularly entrepreneurial streak. In the 1840s, a group of Kulin young men was responsible for the construction of a bridge across the Merri Creek near the recently established Aboriginal school and the Yarra Bend Asylum.[20] The Merri Creek Aboriginal School was one of the earliest institutions in Victoria dedicated to the education of the Kulin people. Begun in 1845, it sought to Christianise and educate as well as enable a sort of self-sufficiency from its onsite vegetable gardens and small number of stock animals. In 1848, new schoolmaster Francis Edgar arrived from Hobart accompanied by his wife, daughter Lucy, and mother-in-law. Lucy wrote one of the earliest firsthand accounts of early Melbourne in a detailed memoir. She spent a lot of time with the young Kulin men, whom she called 'black boys', and noted that prior to 1848 people crossed Merri Creek using a rickety bridge made of gum-tree logs that had been pushed and lashed together. Despite the flimsy handrail that had been erected, many people were reluctant to cross. This presented an opportunity for the Kulin youths based at the school. Lucy wrote that: 'Little Jemmy had earned a good deal [of money] carrying passengers backwards and forwards in our cart, when the creek was not too high.'[21] Heavy summer rains in 1848 saw the log bridge washed away. As a result, the schoolmaster made the decision to supervise the building of a new, durable bridge. On seeing how proud Jemmy was at 'earning' money, Mr Edgar insisted that only the labour of the Merri Creek Aboriginal students would be used, and Lucy described the process:

> [He] called them [the 'black boys'] together, explained the project, and offered them wages at the rate of fourpence per diem for their work at the bridge, provided there was no sulking, and no necessity for driving them to it … their labour was persevering, so earnest. They never were lazy when called to work at the bridge – never sulky, never grumbling. And it must be remembered that this was all extra work; there was the stock to attend to, the harvest to get in, the garden to keep in order, all the same; and it was only in the afternoons they could work out of doors, because of their morning lessons. It was on account of it being extra work that wages were given.[22]

Lucy describes the construction of the bridge and the great pride of the builders. When the bridge opened in November 1849, George Augustus Robinson and his daughters attended.[23] The five builders who remained at the Merri Creek Aboriginal School received payment for their bridge and, in an act we regard as an innovative economic strategy, instituted a toll system for anyone crossing:

> The boys were accustomed, after the completion of the bridge to run out when they saw passengers about to cross it, and demand a toll. They were always alert to their dues; they did not ask any particular sum, but took whatever was offered, and they ran in to show their gains. 'Me got white money this time – him gentleman'; or 'Him only poor fellar – give me penny'. And everyone seemed willing to add his mite [small sum of money] towards remunerating the boys, remarking [on] the excellence of the structure.[24]

The boys who constructed the Merri Creek bridge were not adapting to or merely using the European economic system: they were creating an opportunity to control the movements of the Europeans and in so doing secure for themselves extra resources. The action of the youths standing on the bridge demanding money was not described as begging – on the contrary, it appears that those crossing and the Edgars themselves saw this as a fair exchange. That the youths accepted a sliding scale of payments depending on whether those crossing were 'poor fellars' or 'gentlemen' certainly suggests that the Kulin were astutely engaged in the socio-economics of the situation.

Accessing European goods, clothes and money was a strategic economic concern of the Kulin throughout the 19th century. The assistant protectors, missionaries and other 'humanitarians' often distributed clothes, blankets and rations as a matter of course; their distribution became centralised in 1860 through the establishment of the Central Board Appointed to Watch Over the Interests of the

Aborigines.[25] In many ways this could be regarded as partially the consequence of the Kulin's successful strategy of engaging with the new economy via soliciting and begging.

The establishment of Melbourne and the colonisation of Port Phillip had a devastating and long-lasting effect on the region's Aboriginal communities. Governing the Kulin was an immediate and ongoing concern for the settlers. Despite the establishment of the Aboriginal Protectorate in the 1840s, which was designed to remove them from the city, the Kulin continued to be visible on Melbourne's streets in the mid-1850s and beyond.[26]

At sea

Much has been written about Indigenous participation in the cattle industries. Less well known are the stories of life at sea. Motivation for people to join the whaling and maritime fleets varied. The crews of whaleboats included 'ticket of leave' convicts and others hopeful of overcoming the economic and social disadvantage that was rife in the early colonies. Indigenous people took to the sea for a range of reasons: some sought adventure and escape, or a desire to acquire the goods of the newcomers and the chance of economic gain. There were probably many other compelling reasons to seek a sailor's life that we can only imagine.[27]

In the 19th century, it is known that at least six and most likely many more Tasmanian Aboriginal men went whaling. There is indirect evidence that many Aboriginal people were interested in the whaling industry. It was commonly believed that Aboriginal men tended to be outstanding harpooners since they had expertise throwing spears. The Hobart-based business attracted many men, and one whaleboat in 1839 was described as having an all-Aboriginal crew.[28]

William Lanné was a Tasmanian Aboriginal man whose tragic and grisly death has overshadowed the events of his life. The William Lanné we want to talk about was a whaler, a creative, ingenious, innovative man who used his access to the colonial economy to advocate for his people. Lanné became well known because he was erroneously described by the popular press and journalists as the last Tasmanian Aboriginal man. He had interacted with Europeans for much of his life. With his family he was transported to Wybalenna settlement on Flinders Island in 1842. He received a basic education, learning to read and write via Bible studies. In 1847, he and the forty-six other survivor-inmates on Flinders Island were forcibly removed to Oyster Cove near Hobart. In the early 1850s Lanné went to sea as a whaler when he joined the crews of the whaling ships *Aladdin*, *Jane* and *Runnymede.* In 1860 he was a key harpooner on the colonial whale ship *Sapphire.* On board the *Sapphire* he sailed the Southern Ocean, travelling as far as the Indian Ocean and into the Pacific.

The motivation for Lanné joining the whaling fleet out of Hobart may have included economics and profit-making, and the kind of freedom that life at sea afforded. His life on board, like the lives of his shipmates, was likely organised around tasks and skills: there was no segregation on Australian colonial ships, even in terms of dining. The voyages from Hobart out into the Pacific and the Southern Ocean were relatively brief, and the ships would have been well-provisioned. Evenings were spent relaxing, singing and occasionally dancing, as well as mending sails and other items. Lanné had a great deal of freedom in Hobart's colonial society in contrast to the other Aboriginal people who stayed in the colony. As he was self-sufficient financially, he was free to come and go from Hobart and Oyster Cove. When whalers were in port, they typically stayed at the Dog and Partridge Inn, where Lanné regularly lived and, indeed, where he died.

William Lanné, by virtue of his personality, social skills and independence, exercised an unprecedented degree of personal autonomy and subjectivity. He was committed to bettering the living conditions of those he referred to as 'my people'. According to Lyndall Ryan, he was well aware of his special situation and used it to support better living conditions for the women who remained at Oyster Cove. He wrote to the colonial authorities in 1864: 'I am the last man of my race ... and I must look after my people.'[29] Lanné did not marry or have any children. When he died he was around thirty-four years old; his body was treated with ghastly disrespect and mutilation. His life could easily be consigned to a grisly historical footnote, but we choose to interpret him as someone who seized the opportunities that his changed circumstances offered him and used them for personal and communal gain. His death, while tragic, does not overshadow his agency and capacity. On board the whale ships he found a home, gained acceptance and respect, and sought to make improvements to the lives of 'his people'. In short, he was ultimately innovative, and despite the disruption wrought by colonialism, he found his own ingenious ways of making the most of it.

ECONOMIC ENGAGEMENTS WITH MAKASSANS

We have long been interested in the Makassans and their travels on the seasonal monsoon winds. In October 2022 we had the chance to travel to Makassar, where we met many people whose ancestors were associated with the trepang trade. Until the 1940s this region was known as the Kingdom of Gowa, but today it is called South Sulawesi. Balla Lompoa, a museum dedicated to the history of the Kingdom of Gowa, highlights the relationship between Makassar and Australia. A map on display, drawn to depict Gowa's 19th-century boundaries, includes northern Australia, taking in Arnhem Land and parts of the Gulf of Carpentaria.

Well before Europeans arrived in Australia, mariners from the port of Makassar visited the northern coastline and gathered timber, turtle shell and pearl shell. The visits were stopped in the early 20th century as part of the shameful White Australia policy. In August 2022, the Australian Broadcasting Corporation ran a number of stories about the 'discovery of [an] illegal Indonesian fishing camp'.[30] The periodic visits of mariners from Indonesia is these days met with concern about sovereign borders and biosecurity, but they reflect a tradition that is many hundreds of years old. The continuity of this tradition is elided by journalists and public officials who prefer to tell a tale of an isolated Australia with little contact beyond our shores prior to the arrival of Europeans in the late 18th century. Archaeology and oral histories tell a very different story.

The main reason the Makassans[31] undertook the long journey to northern Australia, which they called Marege, was to collect and process trepang. Trepang is an edible sea cucumber that was traded with the Chinese, who valued it as a delicacy, a medicine and an aphrodisiac. Although many people maintain that there is no documentary evidence of Makassan presence in Marege until the middle of the 18th century, recent work suggests the trade probably dates back significantly further, with sporadic links potentially going back many hundreds of years.[32] Evidence for the Makassans' visits include introduced tamarind trees, the ruins of stone-lined fireplaces where trepang were processed, and Makassan trepang-fishing vessels (*praus*) and other objects depicted in rock art. Linguists have studied the region and have ascertained that there are many introduced Indonesian and Malay words that have become part of the local languages (these are known as loan words or borrow words).[33] Documentary evidence was provided by Matthew Flinders in his circumnavigation of Australia, when he recorded in his journal a meeting in 1803 between his ship *The Investigator* and a fleet of roughly 1000 Makassan trepangers.[34]

Interactions with the seasonal visitors brought new materials and technologies to northern Australian communities. Oral histories and archaeological and anthropological records have shown that Aboriginal communities of north-eastern Australia were highly selective with what they incorporated into their cultures.[35] Well into the historical period, Aboriginal communities continued to trade with their Makassan guests and worked in the Indonesian trepang industry in the 1870s–80s.[36] The Makassan trepangers not only harvested sea cucumbers but also traded turtle shell, pearls, pearl shell and timber (sandalwood).[37] They ensured their *praus* were stocked with trade items including bamboo pipes and tobacco, cloth, belts, string, alcohol, rice and tamarind. The trade lasted for at least five and probably more than nine generations of cross-cultural contact.[38] Aboriginal people also obtained ceramics, glass and metal objects, which they valued highly and often carefully maintained.

Northern Australian communities, particularly Yolŋu, Yanyuwa and the Anindilyakwa of Groote Eylandt, traded with and worked alongside the Makassans. These communities chose to take part in this trade and exchange, cementing relationships and securing access to new materials such as iron, glass and ceramics.[39] There is some indication that this trade had an flow-on impact: Aboriginal people around the year 1800 demanded glass beads when their cooperation and involvement in 'non-customary enterprises' was requested.[40]

One of the most significant technological innovations that accompanied the Makassan exchanges was the introduction of sailing dugout canoes. Makassan traders introduced dugout canoes in the early 18th century, and by the late 1830s and the 1840s the Aboriginal communities were manufacturing these themselves.[41] Anthropologists and linguists working with the Anindilyakwa have documented a great deal of Makassan seafaring terminology and extensive knowledge of *prau* manufacture.[42]

Across northern Australia where the Makassans visited, many rock-art sites include images of Indonesian material culture. In addition to paintings of *prau*s, there are paintings of traditional Indonesian knives, smoking pipes, containers and rope, and even an image of what is thought to be a monkey.[43] Recent work by rock-art expert Sally May and colleagues has revealed that watercraft are

FIGURE 1: Glass plate negative – Fish-tail bark and dug-out canoes. Melville Island, Northern Territory in the background, 1911.

among the most commonly painted images over the past 500 years in some parts of Arnhem Land.[44] The team documented that the overwhelming majority of ship images were of European vessels, with far fewer being Indonesian. This, they argue, is evidence that the Makassans were not seen as usurpers and that they and the northern Australian communities enjoyed harmonious and long-term relationships. May and her colleagues argue persuasively that the plethora of European-ship rock art might be read as statements of sovereignty.[45] In contrast to Europeans, who were figured as invaders, usurpers and colonisers, Makassans were thought of as visitors, allies or trading partners. The few Makassan-ship paintings are regarded as indicative of friendly, negotiated cultural exchanges over a very long period, while the many images of European ships (and horses and guns) are figured to be statements about the permanence of European settlement and imposition.

LYNETTE RUSSELL

In 2021, in between pandemic lockdowns, I had the opportunity to spend a few days in north-east Arnhem Land. I was particularly keen to visit Makassan Beach, also known as Garanhan, 35 kilometres from Yirrkala. Yolŋu friends encouraged me to also visit Bawaka and look out for Bayini, a ghostly pale figure who walks along the beach and is variously described as the haunted spirit of a Makassan woman or a light-skinned spirit figure wandering the shore looking for her kinfolk to return. The late Yolŋu singer Gurrumul[46] wrote and performed in his language of Yolŋu Matha the song 'Bayini', which he described as a 'spiritual love song'. Recently our colleague Lily Yulianti Farid and the record label Skinnyfish remixed the song and incorporated the vocals of a Makassan singer, Dian. Sadly, I never did get to see Bayini when I visited Bawaka. But I will always look for her.

7

REPURPOSING TECHNOLOGIES, ART AND COSMOLOGIES

Innovation can be generated entirely from within society and can also occur in response to new external circumstances. Throughout much of First Nations Australia's 65,000 years of history, different groups have developed their own innovative responses to environmental changes. Responding to the end of the Ice Age and the flooding of the continental shelf and a quarter of the enlarged Australian landmass required massive changes. In other situations, communities became the innovative agents of environmental change, such as with the introduction of landscape burning and aquaculture (Chapter 5). Building on the economic opportunities explored in Chapter 6, here we can examine creative responses of Aboriginal and Torres Strait Islander communities to the arrival of outsiders and their materials and ideas, starting with Makassan visitors from Indonesia and followed up with colonial invasion by the British.

Among the devastating impacts of British invasion, First Nations Australians have found innovative ways of repurposing the material and cosmological offerings of outsiders to enhance their own cultures and societies. In this chapter, we illustrate this repurposing and its obvious and sometimes less obvious consequences. We do this through the lens of innovative and adaptive repurposing of objects and materials for technology and art, and also in relation to cosmologies and religious beliefs from the 18th through to the 21st centuries.

TECHNOLOGICAL TRADITIONS

Metal tools

If you walk into the Ian Potter Centre of the National Gallery of Victoria in Melbourne's Federation Square and head upstairs to the 19th Century Australian Art gallery, the first work you see is an oil painting by Harden Sidney Melville dated 1874 and labelled *Torres Strait canoe and five men at the site of a wreck on the Sir Charles Hardy Islands, off Cape Grenville, North East Australia.*[1] Melville created the painting in England based on memories and sketches he made while he was a young artist aboard the HMS *Fly* during its hydrological survey of Torres Strait for the British Admiralty in the mid-1840s. It shows a group of Torres Strait Islanders in a huge double-outrigger canoe; they have voyaged nearly 200 kilometres south along the Great Barrier Reef and have stopped off at the

Sir Charles Hardy Islands to salvage items that have washed up on the rocky shore from the wreck of a European sailing ship.

Torres Strait Islanders coveted metal for tools and soon learnt that the many European ships that came to grief on the treacherous and often unmapped coral reefs of the region were a great source of metal. Shipwreck salvaging by Torres Strait Islanders during the 19th century was an innovation that probably had its origins back in the 18th century, when Torres Strait and the northern Great Barrier Reef became a shipping route for the emerging colony of New South Wales. A curious tiny fragment of Chinese ceramic material dating to the 1500s and excavated by Ian on Mabuyag in western Torres Strait possibly points to even earlier salvaging.[2]

It is no secret that First Nations Australians immediately took advantage of the introduction of metal to their world by British colonisers. As anthropologist Nic Peterson pointed out, uptake of metal usually occurred where it could be easily substituted for the existing technology. Examples of Aboriginal uptake of metal substitutes for stone implements include spearheads and hafted chisels.[3] Anthropological archaeologist Harry Allen adds that the speed of uptake of metal by different Aboriginal groups was determined less by ideas of perceived technological efficiency and more by concerns of security of supply of metal and ensuring that 'traditional' technologies continued as a backup.[4]

Aboriginal uptake of metal pre-dates British invasion: metal-using Makassan trepang fishers from Indonesia visited Australia's northern shores from at least the 1700s (see Chapter 6).[5] Archaeological excavations of Makassan trepang-processing stations in Arnhem Land have produced a wide array of metal items, including bronze fishhooks, copper coins, steel axe heads, and fragments of iron cauldron.[6] Apart from trading metal with local host communities along the Arnhem Land coast, the introduction of Makassan objects, especially those made of metal, intensified regional exchange networks among Aboriginal communities, especially between coastal and inland groups. It is also likely that Aboriginal people obtained metal objects when they visited Makassar on *praus*. The crew of HMS *Fly*, including Melville, were informed of such visits when they pulled into Port Essington on the Northern Territory coast in 1845. Indeed, they observed a *prau* arrive with an Aboriginal man on board who had travelled to Makassar the previous year.[7]

Archaeological excavations of Aboriginal middens associated with camp sites along the Cobourg Peninsula by Scott Mitchell revealed a major increase in the use of imported inland stone coinciding with the period of Makassan visits.[8] Mitchell argues that this change reflected increased trade between coastal and inland groups due to the desire for Makassan objects such as metal by inland communities (which coastal communities had aplenty) and the desire for prestigious inland stone by coastal communities (which inland communities had aplenty). He adds that the main incentive for increased trade was enhancement of social relationships between coastal and inland groups.

Flaking glass and ceramics

As with metal, all Aboriginal and Torres Strait Islander groups across Australia took advantage of the availability of glass of European origin, mostly in the form of bottles, to manufacture a wide range of cutting and scraping tools. Glass was often seen as a desirable replacement for stone for the manufacture of flaked tools. However, due to its brittleness, glass tended to be restricted to more delicate cutting and scraping activities – nobody would chop down a tree with a broken bottle! In many places, particularly Torres Strait, small chips of bottle glass replaced quartz chips for cutting people's skin for therapeutic and ritual reasons.[9]

For the most part, uptake of glass for tool manufacture and use by Aboriginal peoples living on the colonial frontier went unrecorded by Europeans. In these circumstances, archaeological research can give profoundly new insights into the activities – and the location of camp sites – of Aboriginal peoples living on what Henry Reynolds termed the 'other side of the frontier' and beyond the gaze of Europeans.[10]

An excellent and evocative example of Aboriginal use of glass on the colonial frontier comes from archaeological excavation of an 1840s shepherd's hut near Beaufort, in south-west Victoria, by Nathan Wolski and stone-tool use-wear and residue expert Tom Loy.[11] European archival records briefly document this hut simply as a lonely dot on an old map. Keen to know more about the life of the unnamed shepherd who used the hut, Wolski began excavating around a small pile of rocks in a large grassy paddock, hoping he'd found the collapsed chimney. His detective work was successful and soon pieces of metal, glass and ceramic from objects used by the shepherd began emerging from the sediment. Curiously, some of the glass had been flaked, suggesting an Aboriginal presence. To shed further light on this possible presence, Loy investigated the edges of the glass tools for microscopic traces that might provide clues on potential uses. To his great surprise, the tools had been used for scraping and cutting wood, and for working plant material (probably *Xanthorrhoea* resin and starchy tubers). Remarkably, on one glass tool Loy detected microscopic fibres dyed red to blue, grey and white, probably from the tool user's clothes. Wolski concluded that the glass tools had been made and used by an Aboriginal person or persons, and that the presence of tuber starches strongly pointed to the presence of an Aboriginal woman preparing plant foods. Furthermore, he posited that the male shepherd was probably cohabiting with an Aboriginal woman. Information on this colonial-frontier relationship does not exist in archival records and was only discovered under the microscope.

In some places in Australia, Aboriginal flaking of bottle glass also extended to flaking ceramic telegraph insulators.[12] That is, the ceramic insulators were removed from telegraph poles and flaked with a hammerstone to produce flakes that were then shaped into cutting and scaping tools. In Central Australia, bottles were placed around the base of telegraph poles by government agents in order to dissuade Aboriginal people from climbing the poles and removing the ceramic insulators for tool manufacture.[13] Edward Stirling, anthropologist on the 1894 Horn Expedition to Central Australia, noted that Aboriginal people 'utilise the porcelain insulators for the manufacture of very beautifully made spear-heads; these are not necessarily robbed from the poles for many are broken by lightning and the broken pieces can be picked up in quantities'.[14]

Kimberley points

During at least the last 1500 years, Aboriginal people in the Kimberley region of Western Australia manufactured intricately flaked stone spearheads for hunting, fighting and trade. From the late 19th century, glass and ceramic materials were also used to create these objects, which are known as Kimberley points. Many were manufactured exclusively for trade with European collectors.[15] When the National Museum of Australia opened in 2001, a stunning array of 'stone tools' was exhibited in a backlit display case, including over 100 glass Kimberley points. Glass was a much-sought-after raw material following its introduction, mostly in the form of bottles, as were ceramic telegraph insulators (see previous section). Today, glass and ceramic forms of Kimberley points dominate museum collections. The abundance of Kimberley points in both public and private collections suggests that in the 19th and 20th centuries, the majority were fabricated for the curio and artefact industry rather than for internal Indigenous purposes. According to anthropologist Kim Akerman, it is clear that point manufacturing was the major occupation for men when they were not engaged in hunting or ceremonial events.[16] Archaeologist Rodney Harrison adds:

> Most museum collections that include stone tools also include Kimberley points, despite the fact that the manufacture of

> these points was undertaken by a comparatively small group of Aboriginal people over a very limited period of time and that their manufacture says very little about pre-contact stone artefact manufacture and much more about the rapid and ingenious engagement of Australian Aborigines with global art-culture markets.[17]

Harrison has, in a series of papers, explored the glass and ceramic Kimberley points that were highly desirable to

FIGURE 2: Porcelain Kimberley point.

19th- and 20th-century artefact collectors.[18] He convincingly argues that the proliferation of glass and ceramic points was an innovation by Kimberley Aboriginal people designed to supply an eager colonial market with trade items. The introduction of new source material – glass and ceramic – allowed the master craftspeople to perfect pressure-flaking and produce symmetrical, delicate and visually beautiful points. Even though these are a post-European contact phenomenon, Kimberley points, which Harrison calls 'virtuoso souvenirs', have been portrayed as the pinnacle of Aboriginal ingenuity and material culture.

ARTISTIC TRADITIONS

In Chapter 6 we considered how new motifs and images were incorporated into rock art, and through the work of Sally May and her colleagues we were able to see that some rock art can be understood as statements of sovereignty. The relatively harmonious relationships with the Makassans resulted in proportionally few Makassan-themed images on rock surfaces. That many images of European ships, horses and guns appear on rock surfaces suggests that these newcomers (Europeans) were figured to be a threat, and the rock-art images are assertions of ownership, statements of sovereignty, and testaments to hostility and dispossession. In a typically innovative response to European contact, Aboriginal artists created new mediums and adopted new elements within traditional art forms for practical and decorative, commercial, cultural and social purposes, as explored here.

Pearl shells

Carved and shaped pearl shells were traded across large areas of the Australian continent similarly to baler shells (see Chapter 4).[19] Whereas baler-shell transcontinental trade dates back more than 2000 years, movement of pearl-shell objects across Australia appears to be largely a post-invasion development. Pearl-shell trade objects were ovoid body adornments up to 20 centimetres in length that were used either as pendants/breastplates hung around the neck or pubic covers strapped around the waist. Many feature intricate interlocking linear designs engraved into the surface that were then infilled with red ochre.

Engraved pearl shells originate in the Kimberley region of Western Australia and were traded across two-thirds of Australia and south towards Geraldton, north to Darwin, east to western Queensland, and south-east to Yalata and perhaps the Flinders Ranges of South Australia and Mootwingee in western New South Wales. Mootwingee is 2500 kilometres from the Kimberley. In some cases, pearl-shell objects were traded from South Australia back to Western Australia, travelling over 4000 kilometres in twenty years.[20] Whereas in the past baler-shell objects were walked across the continent, movement of pearl-shell objects was aided by road vehicles, railways and camel trains.

The transcontinental pearl-shell trade was an artistic and social innovation that developed once pearl shells became widely available with the late 19th-century and 20th-century pearling industry in Western Australia. Indeed, anthropologists Kim Akerman and John Stanton note that some pearl-shell objects are made from cultured pearl shells with their telltale remnant artificial blister pearls.[21]

Reckitt's Blue

Most Aboriginal rock paintings are made using red, yellow, black and white pigments, but a few rare examples of blue have been reported. Indeed, one of the earliest European recordings of Aboriginal rock paintings are George Grey's observations of Wandjina paintings in the Kimberley in 1838. Grey mentions that the figures 'each had a very remarkable head-dress, coloured with a deep bright blue'.[22] Re-examination of these paintings in 1947 revealed not 'deep bright blue' but a 'pale blue and green with some yellow about it'.[23] A nearby search located 'blue stones' that were subsequently identified by geologists as glauconite, a form of mica. Although glauconite is typically described as green, when powdered and mixed with water and used as a pigment on rock it dries to a 'pale blue to bluish-green'.[24]

In the 19th century, numerous Aboriginal groups enhanced their colour palette for rock paintings using Reckitt's Laundry Blue. Washing blue or laundry blue was a common household product used for brightening white linens by removing traces of yellow or other discolouration. Laundry blue's main ingredients were synthetic ultramarine and baking soda, wrapped in a muslin bag. The vibrant blue colour appealed to Indigenous artists, who incorporated it into their works on bark, paper and rock surfaces.[25] Reckitt's Blue was introduced to Indigenous artists in the 19th century and its use became widespread in the mid-20th century. Archaeologists Catherine Frieman and Sally May have shown that the adoption of

FIGURE 3: A small packet of 'Reckitt's Blue'.

new materials and techniques into artistic practices did not result in the rejection of previous practices: rather, these materials and techniques enhanced what people were already doing. Reckitt's Blue pigment was used to depict newcomers and their objects but also to illustrate traditional motifs and images.

Modern Aboriginal art continues to use novel and imported materials, colour schemes and gestures, which add to its energy and impact. Reckitt's Blue, acrylic paint and screenprinting technologies are just a few examples of the new materials and methods artists utilise to create narratives that sustain their traditions, strengthen their links to their communities, and keep them connected to their Country. As Frieman and May note, 'In this blend of innovation and tradition, new identities are negotiated, and old stories are reinvigorated for the challenges of the present.'[26]

Ghost nets

'Ghost nets' is a term used to describe lost, discarded or abandoned nylon and polypropylene fishing nets. These synthetic materials do not easily biodegrade, and the nets can continue to float in the ocean for decades, trapping marine life, birds and other animals and damaging coral reefs and seagrass beds. In northern Australia ghost nets have been a major concern for Indigenous coastal communities who care for the sea and rely on their Sea Country for food and other resources and for spiritual practices. Many of the nets originate in South-East Asia. Over the past decade an artistic trend has emerged whereby ghost nets have been used to create powerful three-dimensional sculptural artworks and installations that highlight environmental concerns. Artists have adapted traditional techniques, such as weaving and knotting, and have produced works that comment on the destructive impact of ghost nets and emphasise the importance of sustainable fishing practices and waste management.[27]

The National Maritime Museum in Sydney and the National Museum of Australia (NMA) are two of the major institutions that have profiled these innovative works of art. The NMA acquired its first ghost-net artwork in 2012 and has collaborated with Torres Strait artists from Erub Arts (Darnley Island Arts Centre), an artist cooperative that notes on its home page: 'We are all connected by the oceans of the world, we must all work together towards their protection.'[28] As a result of this collaboration, Erub artists created *Dauma and Garom*, a large-scale sculptural work that was displayed in the exhibition *Garrigarrang: Sea Country*.

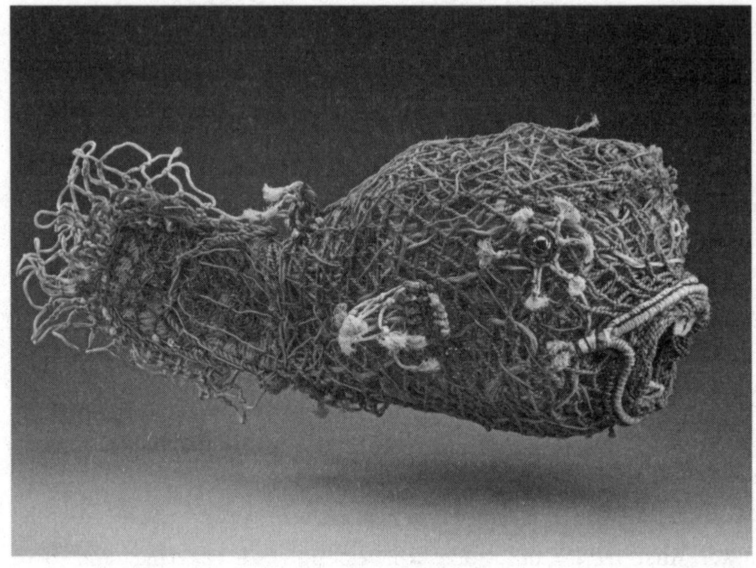

FIGURE 4: *Oyster Cracker* sculpture by Ellarose Savage, 2014.

Ghost-net art is an inventive and creative way to reuse otherwise damaging 'pollution' and at the same time tell the story of environmental concerns and the importance of the oceans for coastal peoples. Each ghost-net artwork depicts a community's struggle to protect its marine environment. Smaller wearable art – jewellery, including earrings and necklaces – and even small woven bags are all raising awareness of the destructive impact of ghost nets.

Fashion

In our home we have a large collection of Indigenous-designed scarves, earrings, necklaces and other accessories. Over the decades we have observed that each year the range of these items grows bigger and the styles more diverse. Some jewellery is woven from traditional basket grasses, while other items are made from ghost nets and other recycled materials. They are admired by both of us, though only worn (mostly!) by Lynette.

Innovative artists such as Maree Clarke have incorporated 3D printing techniques in their work. Clarke matches 3D-printed gold-covered kangaroo teeth and 3D-printed echidna quills in striking statement necklaces, and in 2021 the National Gallery of Victoria exhibited her work in *Maree Clarke: Ancestral Memories.*[29] Lyn-Al Young creates ethereal painted silk garments that are sold in designer stores and have been profiled in *Vogue Australia.*[30] Fashion has long been a vehicle for making statements about heritage and identity. In 2003, Australian Fashion Week launched Dharug fashion designer and painter Robyn Caughlan's ready-to-wear collection. Caughlan went on to have her designs included in the Miss World, Miss Universe and Miss Teen Universe pageants. In May 2023 Wiradjuri designer Denni Francisco, creator of the Ngali label, was the first Indigenous designer to hold a solo show during Australian Fashion Week. First Nations fashion businesses, designers and artists who have worked in the fashion industry for many years had waited a long time for this recognition. Although some Indigenous designers had participated in collaborative fashion shows during the previous two decades, Francisco's solo show was a huge step forward.

Sometimes fashion is not just fashion. Indigenous scholar Treena Clark and design historian Peter McNeil, writing in *The Conversation*, argue that 'First Nations fashion is not just about the catwalk. It is a politically charged practice.'[31] We have often observed Indigenous activists wearing slogan T-shirts or jewellery or other identifiably Indigenous accessories, and we have several Aboriginal friends and family members whose entire wardrobe consists of Indigenous T-shirts or polo shirts. But as we can see from the work of Maree Clarke and Lyn-Al Young, street fashion is not the only place for the communication of Indigenous identity and personal activism. In an academic study of professional Indigenous women's personal style in the workplace, Treena Clark and her colleagues published a sociological study that showed how fashion and attire can serve as an avenue for promoting identity, wellbeing and advocacy. Their study revealed that for many Indigenous women, fashion is an extension of Indigenous identity and an external manifestation of culture. Indigenous fashion and jewellery are recognised as effective communication tools for making a statement about identity and culture.[32]

COSMOLOGIES

Indigenous societies selectively took aspects of newcomers' culture. As we have seen throughout this book, Indigenous cultures in Australia have frequently been additive with regard to external influences. Incorporating elements of newcomers' religion into traditional belief systems can be seen across Australia. Missionaries often thought they had 'converted' their charges to Christianity when what had actually happened was that Christianity and its accompanying rituals were co-opt into the peoples' cosmology.[33]

As Peta Stephenson describes it, Aboriginal 'culture was not destroyed' – rather, 'its infrastructure was extended'.[34]

Christianity

As in most settler-colonial states, European colonisation of Australia was accompanied by Christianity, missions and humanitarians. The history of Christianity's engagement with Aboriginal Australia is lengthy and complex. From the late 18th century onwards, Christian missionaries attempted to convert Indigenous Australians and suppress their traditional religious practices. Many Indigenous communities found ingenious solutions and were able to mesh the new belief systems into their traditional spirituality. When we were undertaking native title research in Victoria we were entertained by some Elders telling us a hilarious story about how their grandmothers, discouraged from using their language, would use traditional words in hymns as a parody. The missionaries thought they were singing songs of the converted but they were, as their granddaughters (now elderly women) told us, 'being cheeky'.

Trawloolway man the Reverend Garry Worete Deverell has reflected on the capacity of many Aboriginal people who blended their traditional beliefs and developed a new syncretic religion and belief system.[35] Somewhat pessimistically, but realistically, Deverell argues that it is impossible to recover in totality a pre-invasion Aboriginality that might then become the primary source for a Christian Aboriginal theology. He notes that it is vital to recover traditional language, history, Law, Songlines and spirituality, but doing so invariably must involve the use of 'post-invasion' resources (linguists, anthropologists) and technology. He laments that Aboriginal 'Indigeneity is, in other words ... forever entwined, both for good and for ill, with the fact of colonialism, with its imagination and its performative effects in hearts, minds, economies and culture'.[36] Aboriginal leaders such as David Unaipon (see Chapter 8) and William Cooper[37] used their Christian faith as a platform to advocate for justice and a fair deal for their people. Today, many Indigenous people have a deep association with the missions, and mission-sites have been protected from development by the communities they once sought to oppress.[38]

Despite the disruptiveness of Christianity, many First Nations communities, or at least individuals, embraced the Bible. Torres Strait Islanders certainly embraced Christian missionisation; indeed, the arrival on 1 July 1871 of the London Missionary Society to the region, specifically Erub (Darnley Island), is celebrated each year with 'Coming of the Light' celebrations. Yet Torres Strait Islander Christianity is unique, blending traditional cosmologies and religious beliefs and practices with Christianity. A number of churches in Torres Strait use giant clam shells as fonts, and St Mary's Church on Mabuyag has the wiwai turtle ritual stone at its entrance (see Chapter 5).

Islam

Indigenous Australians have been acquainted with Islam for several centuries. It was introduced by the Indonesian trepang fishers (see Chapter 6); later, in the 19th century, it arrived with cameleers – that is, camel drivers who mostly came from northern India, Pakistan and Afghanistan. The Indonesian Makassan fishers maintained relationships, particularly with the Yolŋu people of north-eastern Arnhem Land. Evidence for the influence of Islam can be seen in material culture adaptation and in innovative cultural inclusions.[39] Ethnomusicologist Peter Toner identified traces of classical Arabic religious music in Yolŋu songs,[40] and contemporary Yolŋu art often features images of Makassan *praus* (ships), anchors, swords and flags. Ian McIntosh, who lived among the Yolŋu for many years, recorded a song cycle that mentions the law of Walitha'walitha, translated as 'the most high God', or Allah.[41] The song cycle takes place over a number of days and McIntosh describes it as being reminiscent of exchanges with Makassans. It included a flag dance, a knife dance, and narrative elements of Makassar recalled by those Yolŋu who visited the port city. The Yolŋu who visited Makassar also noted seeing shipbuilding, rice paddies and lily ponds.

New animals

The dingo as a placental mammal is not originally native to Australia. Prior to the arrival of Europeans in the 18th and 19th centuries, it was the only introduced animal on the continent. The oldest evidence for dingoes in Australia has been dated to around 3100–3400 years ago.[42] These dates relate to remains found at Madura Cave, on the Nullarbor Plain in the southern half of continental Australia, which indicate that it would have entered from the north earlier than this date. Archaeologists have suggested that the dingo was rapidly adopted by mainland Australian Aboriginal people as a useful companion animal. That is, the dingo was not domesticated but, rather, lived alongside Aboriginal communities, where it was a practical hunting assistant, guardian and protector, and a living blanket or bed warmer.[43] Anthropologist Annette Hamilton suggested that they also kept camp sites clean and sanitised by consuming refuse and faeces, making it possible for people to remain longer in each camp site.[44] According to the ethnographer Deborah Bird Rose, 'The dingo hunts its own food, makes his own camp, finds his own shelter, and follows his own law.'[45] Dingoes were rarely eaten for food and only in times of severe food shortages.[46] They were believed to be able to see mamu (evil spirits), which are invisible to humans, and their barking was interpreted as warding off these spirits.[47]

The dingo became part of the physical and cosmological world of the Aboriginal people of mainland Australia – it never reached Tasmania because it arrived after the flooding of Bass Strait from sea-level rise associated with the end of the Ice Age. Dingoes often feature in Indigenous Australian Dreaming narratives, stories and Songlines. With the probable exception of the snake, the dingo is overrepresented in Indigenous Australian mythology and is associated with the supernatural more than any other animal. It appears in several Dreaming stories throughout the continent.[48] According to Rose, the Yarralin community of the Northern Territory believes that humans are derived from dingoes: the dingo gave birth to humankind and gave humans their characteristic head, genitalia and stance. Humans and dingoes previously were one and the same.[49] The dingo is also often associated with rites of fertility, circumcision and subincision, and is seen as a mediator between the physical and spiritual worlds.[50]

FIGURE 5: Boab nut incised with natural motifs.

In many instances, dingoes were given a burial as one might bury a loved pet. According to rock art expert Ben Gunn and colleagues, a senior Jawoyn man is said to have put dead dingo pups on rock-shelter ledges in southern Arnhem Land as a sign of his affection. In their report of a dingo burial from the early 20th century, they refer to a rock-art image from circa 1920 said to be of a much-appreciated hunting dog. Gunn and others noted that favourite dogs were buried in the same manner as people: wrapped in bark coffins and placed in rock-shelter clefts. A comparable burial of ochred dingo bones was discovered in a shelter in Wardaman Country among human graves, rock art and habitation at a site known as Devil Dingo Dreaming.[51] Dingo burials are more common in south-eastern Australia, where significant cemeteries for people are found; the dingo burials are often associated with these cemeteries.[52]

Beginning in the 18th century, European settlers in Australia introduced a wide variety of animals and plants that radically altered the environment. The introduction of animals with hard hooves, such as cattle, sheep and horses, as well as European-style agriculture, drastically changed fire regimes and altered environments. These changes benefited a small number of local species while harming many others. The most damaging of all introduced animals include rabbits, pigs, goats, deer, camels, water buffalo, donkeys and cane toads. Foxes and feral cats in particular severely harm local wildlife, killing innumerable insects, amphibians, reptiles, small mammals and birds, and taking eggs. Wild horses or brumbies roam alpine regions, and at present Australia has the largest herd of camels anywhere in the world (over 300,000 animals) and the only population of wild camels.

While most of these feral animals are recognised as pests, some have become important food resources and others have been incorporated into local Aboriginal cosmologies, particularly in northern and central Australia. Relationships with introduced animals differ from group to group and region to region. Many Aboriginal people hunted rabbits in south-eastern Australia – in fact, rabbit was often a favoured meat source. The Martu of central Western Australia have for generations hunted cats for food as well as rabbits and, less frequently, camels.[53] While it might be speculated that the hunters would be expected to prefer large animals, such as camels, which would supply vast quantities of meat, they did not.

Instead, the Martu were concerned about the amount of wastage such a large animal would yield.

Innovative programs need to be developed for the management of feral animals on Aboriginal lands, and those programs need to be framed by an appreciation of the values Aboriginal people place on pest animals, especially those that have had a long association with their Country. Both cats and camels are in this category.

Through cosmological inclusion, several animals that might otherwise be regarded as pests have become integrated into Indigenous Dreaming narratives and Songlines. Some introduced animals, such as the horse, have even acquired totemic status.[54] Some groups incorporate donkeys into their religious systems, fusing Christianity with Aboriginal beliefs; because of this they have Donkey Dreamings, as Jesus was known to ride a donkey. Other desert groups maintain Camel Dreamings.[55]

One of the most recorded and studied cosmological innovations is the northern Australian story of the Pussy Cat Dreaming. The Pintupi and Yolŋu, among others, consider cats to be food, not pests. Some researchers have argued that the Pussy Cat Dreaming referred to the native northern quoll or, as it is commonly known, the native cat. Others have recorded that a Yolŋu man was described as the owner of 'Feral Cat Dreaming'. Some Yolŋu artists have incorporated feline elements into their art along with totemic designs related to the artist's clan and moiety.[56] Anthropologist John Bradley has shown that for the Yanyuwa people of Borroloola the pujikat (pussy cat) was readily incorporated into their cosmology and, in fact, the word for cat is the same as the word for quoll: karnbulanyi. A female quoll is called a-kaliba, but this name is not given to cats because introduced animals are always classed as masculine. In a song-cycle study, the old men sang the songs for the quoll, but the quoll for which they sang is not here anymore because 'the white man took it away in a box' and left us with the pussycat. This is how the two animals came to share the same name.[57]

In the southern half of the Northern Territory, the Central Land Council (CLC) manages almost 78 million hectares of remote, harsh and sometimes inaccessible terrain. More than 41.7 million hectares of this area are now owned and managed by Aboriginal people thanks to land rights and native title.[58] The communities try to control the impact of wild animals – such as camels, horses, donkeys and foxes – using both Indigenous knowledges and Western science. However, eradicating feral animals is complicated when those animals have become incorporated into the local economy or cosmology. Aboriginal people often see the use of the land's resources by all animals as natural, and disapprove of killing feral animals if they are going to be wasted and not eaten. According to the CLC's website: 'Many believe that feral animals have earned a right to live on their country through their long association with it, that it's natural for all animals to share their land.'[59]

8

CHANGING AUSTRALIA

The case studies we consider in this chapter explore some of the impacts of Indigenous innovations on broader Australian society. We cover a range of enterprises, from changes to the judiciary system, to land and fire management practices and regimes, to innovative storytelling.

NARRATIVES AND STORIES

Storytelling has always been an integral aspect of First Nations culture. Over at least the past 65,000 years, Indigenous Australians have handed down from generation to generation the stories of ancestors, and chronicles of how the Ancestral Beings created our world. These narratives are often recounted in music, song, dance and art. Each narrative has a distinct purpose and significance: some describe the origins of Country and its characteristics, while others give expression to Law and instructional teachings or societal rules. The stories frequently document change over time. Geographer Patrick Nunn has undertaken long-term studies of oral traditions and Indigenous narratives and has argued that these are often accurate records of Indigenous witnessing. He notes that the stories tell of dramatic changes to landscapes and seascapes, of the Ice Age ending and sea levels rising:

> In some places – Australia, northwest Europe and India – people successfully passed on their observations of coastal drowning for hundreds of generations into the age of literacy and hence down to us today. What science has discovered in the last hundred years or so about postglacial sea-level rise confirms the eyewitness accounts of our ancestors, not the other way round.[1]

Today, many Indigenous people continue the tradition of storytelling, often incorporating digital technology. Since 2011 at Monash University, we have both been involved with Wunungu Awara,[2] a digital storytelling and animation project in partnership with Indigenous communities across Australia that uses animations to help with language preservation and revival. Indigenous communities strengthen their rights to cultural and intellectual property via the preservation and archiving of the history, knowledge, songs and performances of their language. Working under the assumption that

Indigenous communities already have the infrastructure to continue and preserve their language through intergenerational learning, Wunungu Awara supports the existing system by utilising 3D animation as a tool to re-engage and revitalise interest in language continuation by reconnecting language and its people. This innovative and collaborative project uses state-of-the-art technology.

Another dimension of the Monash 3D animation project is bringing to life items of material culture that are known in archival and oral histories but for which no physical examples survive into the present. In 2006, Ian approached Tom Chandler, a 3D animator at Monash University, about the possibility of using historical images as templates to create a virtual Torres Strait canoe. Nineteenth-century paintings, lithographs and photographs indicate that Torres Strait canoes were extraordinary vessels, measuring up to 21 metres in length and fitted with double outriggers that held a central living platform capable of supporting a roofed shelter and fireplace, along with four masts at the bow holding up two large rectangular woven sails. These canoes also featured elaborate carvings, paintings and feather-tassel attachments with ritual significance. Two painted eyes on the bow, and often a woven mouth with shell teeth and fringed with a beard, signified that the vessels were animate object-beings with their own agency and intentionality.[3] No museum in the world holds one of these canoes, probably because Torres Strait Islanders considered their vessels far too valuable to give away or sell to 19th-century collectors.

After years of experimentation, we added another 3D animator, Michael Neylan, to our team and settled on a canoe with attachments minus the sails, as the exact form of the sails remained elusive to us. The final product, fully coloured and textured, was launched and included in the Queensland Museum's 2021–23 *Connections Across the Coral Sea* exhibition (Townsville and Brisbane), where it was projected life-size onto a wall.[4] As the canoe is a 3D animation, it was made to move subtly on the wall, rocking slightly and with its feather tassels gently swaying. Our next step is to take the proof-of-concept canoe on a travelling show to Torres Strait to see if various local island communities would like their own tailor-made 3D canoe animation.

Torres Strait canoes are the largest watercraft known to have been made by First Nations Australians prior to European invasion. They are also the largest portable object of material culture created by First Nations Australians. Beyond representation in European paintings of the late 18th and early 19th centuries, the earliest information on these vessels is rock-art paintings in Torres Strait.[5] The vessels underwrote the specialised maritime culture of Torres Strait Islanders and are thought to date back at least 2500 years.[6] The names of the people who cleverly developed these vessels and subsequently added innovations over the centuries will ever remain with the Old People.

David Unaipon: Inventor and innovator

One Australian Indigenous inventive genius whose name we do know is the polymath and innovator David Unaipon. During his lifetime (1872–1967), Unaipon was described as Australia's best-known Aboriginal man. He was born at the Aborigines' Friends'

Association Mission (AFAM) at Port McLeay (now known as Raukkan) in South Australia, and even as a small child was clearly brilliant. He had musical talent and became a prodigious scientist. Unaipon was a voracious reader with a widely recognised thirst for knowledge. In 1922, the visiting English spiritualist Horace Leaf interviewed him and remarked:

> David is a fine preacher, speaking splendid English, and he is also a good Latin and Greek scholar. According to Mr Leggett [Secretary of the Presbyterian Mission to the Australian Aborigines], he read Newton's *Principia* through and understood even the mathematical equations at the first reading. So capable is he that the Presbyterians proposed sending him shortly on a missionary tour to Tasmania.[7]

Inspired by the manner in which boomerangs spin through the air, Unaipon drew plans for a flying machine that included two boomerang-shaped rotor blades similar to a modern helicopter. His drawings were reported in the newspaper. He explained his design in this 1914 account in the *Daily Herald*: 'An aeroplane can be manufactured that will rise straight into the air from the ground by application of the boomerang principle. The boomerang is shaped to rise in the air according to the velocity with which it is propelled, and so can an aeroplane.'[8] After spending many hours reading the scientific works of Newton and others in the AFAM library, Unaipon became fascinated and excited by the possibility of perpetual motion. His quest led him to invent a new type of mechanical sheep shears.

These were neatly described in the *Adelaide Advertiser* in 1910 under the heading 'An Ingenious Aboriginal':

> The study of mechanics has, it is stated, resulted in the discovery of a new application of force by David Unaipon, ... who has spent a good deal of time at the Point McLeay Mission Station. For five years he attempted to solve the hopeless problem of perpetual motion as applied to machinery, and in the course of his various experiments discovered what he describes as a new method of dealing with the law of gravitation, that is by diverting the attraction to a horizontal instead of a perpendicular movement. He described his discovery to Professor [Robert] Chapman, of the Adelaide University, who advised Unaipon to apply it to machinery. On Monday ... [Unaipon] called at the office and showed how he had adopted the professor's advice. He has altered the mechanism of a machine sheep shears by a device by which the curvilineal motion of the shears is converted into a straight-line movement. At present these machines cut the wool in a half-circular manner, like the motion of the ordinary hand shears, or a pair of scissors. The new mechanism, which is still kept a secret, has been patented, and the inventor states that the principle can be applied to other machinery.[9]

David Unaipon's patented modified handpiece for shearing changed the Australian wool industry because it made shearing much more efficient and resulted in far fewer injuries, such as cuts and nicks, on the sheep. Sadly, Unaipon was not able to pay for a full patent to protect his invention, so he never received credit or financial reward for it. The original provisional patent is available for viewing online at the State Library of South Australia.[10] Over the course of his life, Unaipon provisionally patented many of his inventions, but he never had the resources to pay for the full patent, so his claims lapsed.[11]

Fascinated by tales of ancient Egypt, Unaipon wrote a manuscript on Aboriginal myths and legends that survives today in the Mitchell Library in Sydney. William Ramsay Smith, without acknowledgement of Unaipon's authorship, published these in London as *Myths and Legends of the Australian Aboriginals* (1930). Despite Unaipon's undisputed expertise, his work was appropriated and passed off as that of an Englishman. The book was later republished under his name, but sadly this was decades after he had died.[12]

Unaipon has not been forgotten: his inventions are illustrated on the Australian fifty-dollar note along with his visage. His family did not approve the use of his image and his inventions, and continue to note that the Reserve Bank does not have the appropriate permissions.[13] While this use of the life and work of David Unaipon might be regarded by some as disrespectful, there have been other, more respectful representations. In the following section we discuss some innovative storytelling techniques, starting with Bangarra Dance Theatre's *Unaipon.*

Innovative storytelling

Both in its involvement with Indigenous cultural forms in Australia and in terms of intellectual property, Indigenous dance company

Bangarra Dance Theatre has been creative and innovative. Since 1989 it has sought to tell Aboriginal and Torres Strait Islander stories through unique performances. In 2004 Bangarra produced a work called *Unaipon*, and choreographer Frances Rings was interviewed about it on ABC-TV. She noted that the work was about science, philosophy, writing, the rights of Indigenous people, and religion, and reflected upon the character of David Unaipon:

> I think he was a very tenacious man. He walked in a country that – it was an indomitable landscape. Not many people had … walked that line between the two worlds. It would have been a really difficult time for him. I think he was just blessed with … having a strong spirit.[14]

Another successful theatre company is Ilbijerri, which formed in 1991 in Naarm/Melbourne. Ilbijerri is a local Woiwurrung word meaning 'Coming together for ceremony'. All of the company's performances feature First Nations writers, directors, actors, key creatives and theatre practitioners. Iljiberri tours regional areas and country towns using theatre to tell important Indigenous stories. A celebrated cast member was Uncle Jack Charles, a cherished Elder whom we sadly lost in 2022. Uncle Jack was one of Australia's best-loved Aboriginal performers, a stage and screen actor who lived with a long-term heroin addiction that he sustained by cat-burgling (which he described as getting whitefellas to 'pay the rent'). In 2010, after forty years of 'chasing the dragon'[15] and over twenty years of imprisonment, Uncle Jack (and playwright John Romeril) wrote a highly successful play, *Jack Charles v The Crown*. It was performed for over eight years with Uncle Jack in the lead, and successfully toured nationally and internationally. Ilbijerri calls it 'a much-loved classic' in its repertoire.

Uncle Jack's story was one of Stolen Generations, institutional and sexual abuse, and neglect. He argued that his life of drug dependency and crime stemmed from his early experiences of being removed from his mother and placed in a children's home. Jack was committed to truth-telling and was the first Elder to give evidence at the Yoorrook Justice Commission's public hearings, or wurrek tyerrang. The first of its kind in Australia, Yoorrook is a truth-telling commission that intends to document 'injustices experienced by Traditional Owners and First Peoples in Victoria in all areas of life since colonisation'. Wurruk tyerrang is a western Victorian Wergaia term for 'speaking together'.[16] *Jack Charles v The Crown* is a harrowing story documenting violence, twenty-two incarcerations and extensive drug abuse, stemming from a foundation of systemic oppression. Sadly, Uncle Jack's story is not unique but stands as an example of the impact of child removal and family dislocation. Later in his life Uncle Jack worked closely with Indigenous youth, particularly those who were at risk of being or had been incarcerated.

INNOVATIVE JUSTICE

As in Uncle Jack's case, most Indigenous offending is property-related, the consequences of poverty, or related to needs associated with addiction. Poverty and substance abuse issues have led to disproportionate levels of incarceration of Aboriginal and Torres Strait Islander people. Indigenous Australians, like their counterparts in Canada, New Zealand and the United States of America, are overrepresented in the criminal justice system compared to the general population.[17] Overrepresentation is a complex issue with roots in historical and ongoing systemic discrimination, intergenerational poverty, and the consequences of diminished educational opportunities.

Another major factor has been disconnection from culture, family and Country. The removal of children from their families has been an ongoing blight in Australia, and the Stolen Generations were highlighted in the Royal Commission into Aboriginal Deaths in Custody (1987–91) and the 1997 report published by the Human Rights and Equal Opportunity Commission (HREOC; now the Australian Human Rights Commission) called *Bringing Them Home*, which arose from the National Inquiry into the Separation of Aboriginal and Torres Strait Islander Children from Their Families.[18]

The HREOC report provided evidence of the widespread and systematic removal of Indigenous children from their families between 1910 and 1970, and the children's placement in institutions or with non-Indigenous families. It argued that these acts of removal were a flagrant breach of human rights and highlighted the disastrous effects that government policies had on Indigenous families and communities. The report's publication was a pivotal point in Australian history. In many ways the report has provided a framework for policymakers and the justice system over the past two decades. The federal parliament offered an apology in 2008 to acknowledge the suffering of the removed generations. Sadly, too many Indigenous people continue to be incarcerated and far too many die in custody.

Almost from the very beginning of European colonisation of Australia, Aboriginal people suffered at the hands of the law. In 1835, John Batman and John Pascoe Fawkner arrived from Tasmania to establish the village that became the city of Melbourne. Traffic back and forth across Bass Strait meant that the colonisations of Port Phillip and Tasmania were intimately connected. The control and administration of Aboriginal people in the Port Phillip District were based on experience gained in Tasmania. In charge was George Augustus Robinson, the so-called pacifier of the Tasmanians. On his arrival in Victoria, Robinson was accompanied by a group of Tasmanian Aboriginal people; they had been his fellow travellers for much of the previous decade. Tragically, two of the men, Tunnerminnerwait (known as Jack) and Maulboyheener (known as Bob), became the first people executed in the Port Phillip District under British law. This took place in 1842, a mere seven years after Batman's fraudulent treaty with the Kulin people.[19] Tunnerminnerwait and Maulboyheener were not given a fair trial: their evidence was not heard in court, and their advocate was literally locked out of the proceedings. As Kate Auty and Lynette Russell show, this unjust and unfair trial and execution was not conducted according to the standards of the day. From the early days of the colony, Aboriginal Victorians have been vastly more likely than people from mainstream white society to be arrested, charged and jailed.

It is unsurprising that the emergence of the innovative Koori Courts in Victoria was intimately connected to the HREOC report and the Royal Commission into Aboriginal Deaths in Custody. As Indigenous people seek ways to deal with the judicial system, there have been some very innovative attempts at finding mechanisms that deliver culturally appropriate sanctions. The Koori Courts are designed to address the overrepresentation of Indigenous Australians in the criminal justice system, and to provide responsive justice that takes into account the unique social and cultural circumstances and needs of Indigenous defendants.[20]

Based on the concept of therapeutic jurisprudence, the Koori Courts are intended to reduce recidivism and diminish the potential for harmful consequences that frequently accompany incarceration.[21] They operate as part of the regular court system but have a particular focus on restorative justice. This means that in addition to considering traditional legal penalties, they may consider the impact of a person's actions on their community and may look for ways to repair the harm caused and restore relationships. The staff, a mix of Indigenous and non-Indigenous judges and magistrates, work closely with Indigenous community organisations and support services to provide defendants with access to the help they need to address the underlying issues that may have contributed to their offending. It is important to note that the Koori Courts only hear guilty pleas.[22]

Incorporating Indigenous cultures and traditions into the criminal justice system produces more effective and culturally acceptable responses to Indigenous offending. The Koori Courts seek to advance a more fair and equitable system for Indigenous offenders by emphasising the role of cultural identity and community ties in addressing criminal behaviour. Since the early 2000s they have been successfully negotiating this difficult terrain with outstanding results, including significant reduction in subsequent reoffending.[23]

INNOVATIVE BURNING AND LAND/SEA MANAGEMENT PRACTICES

Much has been written about Australia's firestick farming and the emergence of fire-dependent landscapes. Bill Gammage and Bruce Pascoe in their book *Country: Future Fire, Future Farming* and Zena Cumpston, Michael-Shawn Fletcher and Lesley Head in *Plants: Past, Present and Future* show at length how fire has changed, shaped and created our unique landscapes. It is not our aim here to rehash this information but, rather, to reflect on some innovative recent burning and land management practices.

Aboriginal burning practices have recently been reinstated in northern and central Australia. In particular, strategic early dry-season burning returns these landscapes to the patchiness and mosaic of vegetation seen prior to European invasion.[24] Aboriginal burning can benefit Indigenous communities economically through enhancing the habitat for local animals as well as through the commercial value of carbon sequestration and abatement. In Australia's tropical savannah lands that are prone to fire, Indigenous fire managers are involved in the Western Arnhem Land Fire Abatement project, a partnership between Indigenous communities and the Cooperative Research Centre for Tropical Savannas Management. The project focuses on early dry-season patch burning to prevent wholesale burning: the latter destroys animal habitat and releases significant amounts of carbon into the atmosphere. As part of the partnership, native species of plants and animals, many of which are exclusive to the area, are being meticulously tracked and surveyed.[25] This innovative carbon capture program is now being trialled across other parts of mainland Australia and in Tasmania.[26]

Since the land rights movements of the 1970s and the *Native Title Act* of 1993, Indigenous native title ownership has been formally recognised and significant parcels of land have returned to Indigenous stewardship and management, primarily in rural and remote locations.[27] This reinstatement of Indigenous curatorship of Country initially caused tension as Indigenous land use was perceived as potentially countering environmental and conservation efforts, with non-Indigenous land managers concerned that areas set aside primarily for the protection of biodiversity would be incompatible with Aboriginal people's interests in hunting and resource collection. However, over the past few decades it has become clear that shared management of national and state parks and reserves is a powerful and effective strategy for incorporating Indigenous knowledge. Innovative solutions have been found that allow hunting, including the use of guns and vehicles, to be compatible with the preservation of biodiversity.[28] This process marked the beginning of a gradual shift towards including Indigenous peoples and their knowledge in land management. The change was prompted in part by the acknowledgement of First Nations' rights and of the high biodiversity and other environmental assets of their territories. At the same time, Aboriginal communities all over Australia were reviving their ancient traditions and customs and, in some cases, returning to their ancestral lands.

As a result of the changes in environmental policy and practice, Indigenous Australians are now involved in various natural resource management projects, including scientific research. Indigenous rangers are working to control feral animals, monitor endangered species, and keep scientific records. Many are supported by initiatives centred on Torres Strait and the north of the continent, where over 45 per cent of the land is managed by Indigenous people. The Indigenous Protected Areas (IPA) program of the Australian Government, which was born out of native title developments, offers a framework for the management of these Indigenous-owned properties as a part of the National Reserve System.[29]

The successful Indigenous ranger Working on Country program was created by the federal government in 2007 to provide extensive training and career opportunities for Indigenous people in land and sea management. It combines Indigenous understandings of landscapes and seascapes, flora and fauna, with Western scientific training, and has resulted in the establishment of meaningful on-Country jobs.[30] Indigenous ranger teams collaborate with ecological and conservation researchers and a broad range of philanthropic, economic and educational organisations. Elders across northern Australia, in particular, have worked alongside government and non-government agencies, schools and community organisations to promote intergenerational teaching and skill sharing.

Indigenous people's involvement has led to sustainable jobs in the environmental, biosecurity, tourist, heritage and other wildlife-related industries.

The North Australian Indigenous Land and Sea Management Alliance (NAILSMA) is an Indigenous-owned and Indigenous-controlled organisation that was set up to support traditional owner groups across northern Australia. According to the NAILSMA website, its programs combine Indigenous knowledge and contemporary Western scientific expertise. Its special focus is on maintaining and managing faunal populations on land and in the sea. Various Indigenous community organisations, traditional owner groups, Indigenous rangers, regional organisations, Aboriginal land councils, native title holders and academic researchers are involved in its extensive projects. Because of its emphasis on conservation, NAILSMA was the first Indigenous-led organisation in Australia to join the International Union for Conservation of Nature (IUCN).

NAILSMA has been intensively involved with dugong and sea turtle research. Right across northern Australia, coastal Indigenous groups maintain their social, cultural and spiritual ties with these marine animals and collaborate closely with government organisations, including NAILSMA, to manage the land and sea in line with continued customary use. Indigenous communities continue to hunt dugongs and all six species of marine turtle found in Australia, including the olive ridley, leatherback, green, flatback and hawksbill turtles. All of these are threatened with extinction and, as a result, are protected by international accords such as the Bonn Convention (Convention on the Conservation of Migratory Species of Wild Animals) and are designated as endangered or vulnerable under the Australian Government's *Environment Protection and Biodiversity Conservation Act* 1999 and the IUCN Red List of Threatened Species. However, through innovative strategies, Indigenous communities are ensuring that traditional hunting and consumption of these meats is sustainable.[31]

The Yirralka Rangers program associated with the Laynhapuy IPA, on Yolŋu Country at Nhulunbuy on the Gove Peninsula, Northern Territory, is one of the coastal ranger organisations monitoring sea turtles to ensure that the hunting of these animals and collection of their eggs does not have an adverse effect on population numbers.[32] These highly trained rangers monitor the annual turtle catch and count nests with eggs. They measure the turtles' dimensions, record their sex, and note their condition and age, with their findings helping to map out population trends. All of this activity is motivated by a desire to ensure sustainable hunting and advance scientific research.

In addition to turtle research, the Yirralka Rangers – and other organisations, such as the Li-Anthawirriyarra Sea Rangers on Yanyuwa Country in the Gulf of Carpentaria – are active in a range of sea and land activities, including weed management, burning regimes and feral animal removal. On Sea Country these organisations manage removing marine debris, overseeing fishing operations, monitoring fish populations, conducting research on marine biodiversity, and restoring the habitats of vulnerable and endangered species.[33]

An innovative approach to dugong management was initiated by the Goemulgal community of Mabuyag of central-western Torres

Strait, who wished to use archaeological evidence to demonstrate the long-term sustainability of their dugong-hunting practices. Twenty years ago, a series of academic papers presented detailed survey data on dugong numbers in Torres Strait to argue that the rates of hunting by Torres Strait Islanders were unsustainable, most likely due to modern hunting equipment and practices. To the Goemulgal, who were among the major hunters of dugongs in the region, the conclusion of unsustainability did not gel with their understanding of the practice and what they saw as maintenance of 'traditional' hunting rates. As a result, they were interested to know if ancient dugong bones they had excavated with Ian from ritual dugong bone mounds and ancestral village sites on Mabuyag could provide data on precolonial hunting rates and hunting sustainability. The data from dugong bones in the Dhabangay bone mound suggested that the number of dugongs hunted in the 200–300 years leading up to colonial settlement in the 1870s was similar to numbers hunted in the present.[34] Subsequent measurement of the size of dugong ear bones (an indicator of the size and age of the animals) in mounds and middens revealed extraordinary consistency in the hunting of adult dugongs over the past 1000 years, with no evidence of unsustainable hunting rates.[35] Presented with this new evidence, conservation biologists reconsidered their data and findings on unsustainability, and dugong hunting in Torres Strait is now considered sustainable under the stewardship of Torres Strait Islanders.[36]

9

LOOKING FORWARD

In this book we have strived to show that innovation and inventiveness have been at the core of First Nations cultures for at least 65,000 years. We have touched upon only a fraction of the examples that could have been used to illustrate this innovation and inventiveness. In terms of stories from before European invasion, First Nations communities working closely with archaeologists have barely scratched the surface when it comes to understanding our ancient history. Although legendary and mythological Dreaming narratives, along with oral histories that extend back thousands of years, already populate Australia's ancient history, tangible legacies of the Old People in the form of archaeological sites and objects provide new opportunities for younger generations of First Nations peoples to be part of the process of elaborating on and adding to those old narratives with their own new narratives of their ancestors' achievements.

But archaeological approaches to our ancient history must first resonate with First Nations Australians before they can have currency with the rest of Australian society. In the past, Australian archaeology sang from the earlier score sheet of spurious anthropological theories that did a disservice to First Nations peoples with notions of primitivism, conservatism and even uninventiveness.[1] New generations of archaeologists have worked hard to separate themselves from what can be regarded as a somewhat shameful disciplinary history. Credibility in this regard comes through action, and listening to and learning from First Nations communities with whom we work. We are on the right track and Australia is in many respects at the forefront of global innovation when it comes to Indigenous archaeology.[2]

Historians, too, have adjusted their methods and research practices. Projects that involve Indigenous histories must include Indigenous voices. Through our decades of teaching, supervising graduate students, writing, researching and conducting public outreach, we have both endeavoured to ensure that the work we have been involved in adheres to the maxim 'Nothing about us without us'.[3] Each of us has had stints as president of our respective professional associations. Ian was president of the Australian Archaeological Association Inc. 2007–09; Lynette was on the executive of the Australian Historical Association from 2014 to 2018, serving as its president 2016–18. These roles enabled us to advocate effectively for disciplinary change, and in particular to seek out and listen to Indigenous people and their critiques.

If ever there was a time to listen, it is now. As we have been writing this book, the debate about a Voice to Parliament has swirled around us. The referendum that will take place in October 2023 is the product of thousands of hours of community engagement, grassroots meetings and the work of dedicated activists. In 2017, Indigenous leaders drafted the Statement from the Heart, often known as the Uluṟu Statement. It advocates for a 'First Nations Voice' to be established in the Australian constitution and for a Makarraṯa Commission to oversee an agreement-making process between Indigenous and non-Indigenous Australians. As Indigenous legal scholar Asmi Woods explains, makarraṯa is a Yolŋu word for 'a negotiation of peace' or 'peace after a dispute'.[4]

Like many, we were disappointed when a previous government dismissed this idea out of hand with what appeared to be minimal consideration. We are hopeful that the referendum will pass and a Voice to Parliament becomes a reality. It is an innovative solution that will ensure Indigenous people have a voice and are heard on matters relevant to them. As we have shown throughout this text, we believe that Australian Indigenous cultures have been innovators par excellence for millennia. We have no doubt that a people who have found ingenious ways of living, surviving and thriving on this continent will find innovative ways of dealing with contemporary challenges.

The idea of a voice and being heard drags us back to the 1968 Boyer lectures, when anthropologist WEH Stanner accused white Australia of 'a cult of forgetfulness practised on a national scale'.[5] Stanner claimed that history and culture of Indigenous Australians had been largely ignored and forgotten by the wider Australian society. He called this forgetfulness the 'Great Australian Silence'. Aboriginal people have occupied this country for at least 65,000 years and witnessed climatic changes we can only marvel at, but they found ways of thriving in an often-changing landscape. Through an Ice Age with large-scale droughts and shifting sea levels, they maintained their networks and connections and developed their culture. Their feats are recorded in their artworks, their camp sites and their mortuary practices and burials. Yet the historical forgetting was an intentional erasing tactic by which Indigenous people and their culture were ignored, excluded and suppressed. According to Stanner, this process of forgetting was not unintended or incidental but, rather, an intentional colonial policy. The tragedy of that silence is that many wondrous cultural achievements were denied to generations of Australians. We were denied the history of our own country.

As we move forward, thanks to the research featured in the First Knowledges series, we are optimistic that we are on the threshold of a new era. We hope that Australia as a nation celebrates its extraordinary Indigenous history and takes pride in the innovations of the past and those that will undoubtedly emerge in the future.

ACKNOWLEDGEMENTS

We would like to thank the Australian Research Council Centre of Excellence for Australian Biodiversity and Heritage for support. We also thank Monash University for its continued support of Indigenous studies. To our Monash University colleagues we say a big thank you, and Lynette particularly thanks the Global Encounters team. Many of our views have benefited greatly from discussions with academic colleagues. In particular we single out anthropologist John Bradley, who has helped open our eyes to new worlds of understanding.

IMAGE CREDITS

Front inside cover

A clear green glass Kimberley point or spearhead, bifacially flaked with serrated edges. It is rounded at the proximal end and tapers to a point. Handwritten in white on one side is 'A-WH 147'.
Creator: Unknown
Photo: Jason McCarthy
Image courtesy of National Museum of Australia

Back inside cover

Sea turtle sculpture by Christine Yantumba, 2019
A sculpture of a turtle made from ghost nets, plastic trellis, zipties and twine. It is painted yellow, green and red with acrylic and fluorescent paint. There are circles on the top of the shell and wavy lines on the flippers.
Creator: Christine Yantumba, Pormpuraaw Arts
Photo: Jason McCarthy
Image courtesy of National Museum of Australia
© Christine Yantumba/Copyright Agency, 2023

129

Glass plate negative – Fish-tail bark and dug-out canoes. Melville Island, Northern Territory in the background, 1911.
Creator: Herbert Basedow
Reprography: Lannon Harley
Image courtesy of National Museum of Australia

138 A white porcelain Kimberley point with a broad proximal end tapering to a sharp point at the distal end; the surfaces are flaked, and the edges serrated. Handwritten in black on one side is 'Kimberley/W.A.'.
Creator: Unknown
Photo: Dragi Markovic
Image courtesy of National Museum of Australia

142 A small packet of 'Reckitt's Blue'.
Photo: Jason McCarthy
Image courtesy of National Museum of Australia

144 *Oyster Cracker* sculpture by Ellarose Savage, 2014
A round sculpture of a fish made from ghost nets (nylon monofilament fishing lines) with an open mouth and teeth. The fish is blue, purple, orange, yellow, and turquoise in colour with a yellow plastic frame. The fish has two shiny plastic eyes.
Creator: Ellarose Savage, Erub Arts
Photo: Jason McCarthy
Image courtesy of National Museum of Australia
© Ellarose Savage/Copyright Agency, 2023

151 Boab nut incised with natural motifs.
Creator: Unknown
Photo: Amanda Crnkovic
Image courtesy of National Museum of Australia

NOTES

1. PERSONAL PERSPECTIVES

1 See Donald Barr, *The How and Why Wonder Book of Primitive Man* (Wonder Books, New York, 1961).

2 Editors of Life, *The Epic of Man* (Time Incorporated, New York, 1961).

3 Editors of Life, *The Epic of Man*, pp. 242–7.

4 Our first research collaborations were Ian J McNiven & Lynette Russell, 'Place with a Past: Reconciling Wilderness and the Aboriginal Past in World Heritage Areas', *Royal Historical Society of Queensland Journal*, 15, 1995, pp. 505–19; Ian J McNiven, Lynette Russell & Kay Schaffer (eds), *Constructions of Colonialism: Perspectives on Eliza Fraser's Shipwreck* (Leicester University Press, London, 1998); and Lynette Russell & Ian J McNiven, 'Monumental Colonialism: Megaliths and the Appropriation of Australia's Aboriginal Past', *Journal of Material Culture*, 3, 1998, pp. 283–99.

5 Ian J McNiven & Lynette Russell, *Appropriated Pasts: Indigenous Peoples and the Colonial Culture of Archaeology* (AltaMira Press, Lanham, MD, 2005); Ian J McNiven & Lynette Russell, '"Strange Paintings" and "Mystery Races": Kimberley Rock Art, Diffusionism and Colonialist Constructions of Australia's Aboriginal Past', *Antiquity*, 71(274), 1997, pp. 801–9; Ian J McNiven, 'The Bradshaw Debate: Lessons Learned from Critiquing Colonialist Interpretations of Gwion Gwion Rock Paintings of the Kimberley, Western Australia', *Australian Archaeology*, 72, 2011, pp. 35–44.

6 Ian J McNiven, 'The Ethnographic Echo: Archaeological Approaches to Writing Long-Term Histories of Indigenous Spiritual Beliefs and Ritual Practices', *Humanities Australia*, 7, 2016, pp. 8–21. An example of our co-designed collaborative research is Ian J McNiven, Bruno David, Goemulgau Kod & Judith Fitzpatrick, 'The Great *Kod* of Pulu: Mutual Historical Emergence of Ceremonial Sites and Social Groups in Torres Strait, NE Australia', *Cambridge Archaeological Journal*, 19(3), 2009, pp. 291–317.

7 The Australian Institute of Aboriginal and Torres Strait Islander Studies (AIATSIS) website notes that there were 250 Indigenous languages, including up to 800 dialects, and perhaps 600 different groups, who are often referred to as 'tribes'. AIATSIS is a collecting and research agency funded and sanctioned by the federal government and dedicated to Indigenous Australia. Lynette Russell has been a member of AIATSIS for over twenty years.

8 Lynette Russell published the story of her search in *A Little Bird Told Me* (Allen & Unwin, Crows Nest, NSW, 2002).

9 Lynette Russell, *Savage Imaginings: Historical and Contemporary Constructions of Australian Aboriginalities* (Australian Scholarly Publications, Melbourne, 2001), p. 3.

10 Billy Griffiths & Lynette Russell, 'What We Were Told: Responses to 65,000 Years of Aboriginal History', *Aboriginal History*, 42, 2018, p. 54. For an up-to-date overview of the 65,000-year archaeological story of First Nations Australia, see Ian J McNiven & Bruno David (eds), *The Oxford Handbook of the Archaeology of Indigenous Australia and New Guinea* (Oxford: Oxford University Press, 2023).

11 See Russell, *Savage Imaginings*.

12 McNiven & Russell, *Appropriated Pasts*, Chapter 3.

13 Fred D McCarthy, 'Aboriginal Australian Material Culture: Causative Factors in Its Composition', *Mankind*, 2(8), 1940, pp. 241–69 and 2(9), 1940, pp. 294–320. For an analysis and discussion of the diffusionist paradigm in Australian archaeology, see Ian J McNiven, 'Colonial Diffusionism and the Archaeology of External Influences on Aboriginal Culture', in Bruno David, Bryce Barker & Ian J McNiven (eds), *The Social Archaeology of Australian Indigenous Societies* (Aboriginal Studies Press, Canberra, 2006), pp. 85–106.

14 Herbert VV Noone, 'Some Aboriginal Stone Implements of Western Australia', *Records of the Australian Museum*, 7(3), 1943, pp. 271–80.

15 Noone, 'Some Aboriginal Stone Implements of Western Australia', p. 279.

16 Fred D McCarthy, 'The Use of Stone Tools to Map Patterns of Diffusion', in Richard VS Wright (ed.), *Stone Tools as Cultural Markers:*

Change, Evolution and Complexity, Australian Institute of Aboriginal Studies Prehistory and Material Culture Series No. 12 (Humanities Press, Atlantic Highlands, NJ, 1977), pp. 251–62.

17 Grafton Elliot Smith, *Human History* (Jonathon Cape, London, 1930), p. 262. Hyperdiffusion is the label given by anthropologists and archaeologists to a now-discredited school of thought that attributed the origin of many cultural practices across the world to a single advanced ancient culture, usually Egypt. Although cultures borrow ideas from neighbouring groups, archaeological research has shown that nearly all of the world's cultural practices attributed to ancient Egypt by hyperdiffusionists were local developments. In many respects, hyperdiffusionism had racist overtones as it denied the capacity of local innovation by Indigenous peoples. For a detailed critique of hyperdiffusionist interpretations of Australian Indigenous cultural practices, see McNiven & Russell, *Appropriated Pasts*, Chapter 5.

18 Rev. Peter MacPherson, 'The Oven-Mounds of the Aborigines of Victoria', *Journal and Proceedings of the Royal Society of New South Wales*, 18, 1885, pp. 49–59. For a discussion of MacPherson's ideas, see Ian J McNiven, '"The Thick Darkness of Pre-Historic Times": Antiquarian Archaeology in Nineteenth-Century Victoria', in McNiven & David (eds), *The Oxford Handbook of the Archaeology of Indigenous Australia and New Guinea*.

19 Herbert M Hale & Norman B Tindale, 'Notes on Some Human Remains in the Lower Murray Valley, South Australia', *Records of the South Australian Museum*, 4(2), 1930, pp. 145–218.

20 Griffiths & Russell, 'What We Were Told', pp. 31–53.

21 Harry Lourandos, 'Change or Stability? Hydraulics, Hunter-Gatherers and Population in Temperate Australia', *World Archaeology*, 11(3), 1980, pp. 245–66; Harry Lourandos, 'Intensification: A Late Pleistocene-Holocene Archaeological Sequence from Southwestern Victoria', *Archaeology in Oceania*, 18, 1983, pp. 81–94.

22 For a detailed exploration of the theories and legacies of Harry Lourandos, see various chapters in David, Barker & McNiven (eds), *The Social Archaeology of Australian Indigenous Societies*.

23 For an up-to-date synthesis of current theoretical views on the archaeology of Indigenous Australia, see McNiven & David (eds), *The Oxford Handbook of the Archaeology of Indigenous Australia and New Guinea.*

24 JP White & DJ Mulvaney, 'How Many People?', in D John Mulvaney & J Peter White (eds), *Australians to 1788* (Fairfax, Syme & Weldon, Broadway, NSW, 1987), pp. 115–17; Corey JA Bradshaw et al., 'Stochastic Models Support Rapid Peopling of Late Pleistocene Sahul', *Nature Communications*, 12(1), 2021, p. 2440.

25 Natalie R Franklin & Phillip J Habgood, 'Modern Human Behaviour and Pleistocene Sahul in Review', *Australian Archaeology*, 65(1), 2007, pp. 1–16. See also Phillip J Habgood & Natalie R Franklin, 'The Revolution that Didn't Arrive: A Review of Pleistocene Sahul', *Journal of Human Evolution*, 55(2), 2008, pp. 187–222.

26 Cited in Franz-Josef Micha, 'Trade and Change in Australian Aboriginal Cultures: Australian Aboriginal Trade as an Expression of Close Culture Contact and as a Mediator of Culture Change', in Arnold R Pilling & Richard A Waterman (eds), *Diprotodon to Detribalization: Studies of Change Among Australian Aborigines* (East Lansing: Michigan State University Press, 1970), pp. 292–303.

27 Cited in Micha, 'Trade and Change in Australian Aboriginal Cultures', p. 292.

28 Micha, 'Trade and Change in Australian Aboriginal Cultures', p. 293. Original emphasis.

2. ICE AGE TECHNOLOGIES OF SEA AND LAND

1 Chris Clarkson et al., 'Human Occupation of Northern Australia by 65,000 Years Ago', *Nature*, 547(7663), 2017, pp. 306–10; Lynda M Petherick et al., 'An Extended Last Glacial Maximum in the Southern Hemisphere: A Contribution to the SHeMax Project', *Earth-Science Reviews*, 231, 2022; Yusuke Yokoyama et al., 'Sea-Level at the Last Glacial Maximum: Evidence from Northwestern Australia to Constrain Ice Volumes for Oxygen Isotope Stage 2', *Palaeogeography, Palaeoclimatology, Palaeoecology*, 165(3–4), 2001, pp. 281–97.

2 Josephine Flood, *Archaeology of the Dreamtime* (Collins, Sydney, 1983); J Peter White & James F O'Connell, *A Prehistory of Australia, New Guinea and Sahul* (Academic Press, Sydney, 1982).
3 John H Calaby, 'Some Biological Factors Relevant to the Pleistocene Movement of Man in Australia', in RL Kirk & AG Thorne (eds), *The Origin of the Australians* (Australian Institute of Aboriginal Studies, Canberra, 1976), pp. 23–8.
4 Joseph B Birdsell, 'The Recalibration of a Paradigm for the First Peopling of Sahul', in Jim Allen, Jack Golson & Rhys Jones (eds), *Sunda and Sahul: Prehistoric Studies in Southeast Asia, Melanesia and Australia* (Academic Press, London, 1977), pp. 113–67.
5 Noel Butlin, *Economics and the Dreamtime: A Hypothetical History* (Cambridge University Press, Cambridge, 1993); Shimona Kealy, Julien Louys & Sue O'Connor, 'Islands Under the Sea: A Review of Early Modern Human Dispersal Routes and Migration Hypotheses Through Wallacea', *Journal of Island and Coastal Archaeology*, 11(3), 2016, pp. 364–84; Kasih Norman et al., 'An Early Colonisation Pathway into Northwest Australia 70,000–60,000 Years Ago', *Quaternary Science Reviews*, 180, 2018, pp. 229–39; Kasih Norman, Sue O'Connor & Michael Bird, 'Island Hopping to Sahul', in McNiven & David (eds), *The Oxford Handbook of the Archaeology of Indigenous Australia and New Guinea*; Michael I Bird et al., 'Palaeogeography and Voyage Modeling Indicates Early Human Colonization of Australia Was Likely from Timor-Roti', *Quaternary Science Reviews*, 191, 2018, pp. 431–9.
6 Corey JA Bradshaw et al., 'Minimum Founding Populations for the First Peopling of Sahul', *Nature Ecology & Evolution*, 3(7), 2019, pp. 1057–63.
7 Nano Nagle et al., 'Mitochondrial DNA Diversity of Present-Day Aboriginal Australians and Implications for Human Evolution in Oceania', *Journal of Human Genetics*, 62(3), 2017, pp. 343–53; Ray Tobler et al., 'Aboriginal Mitogenomes Reveal 50,000 Years of Regionalism in Australia', *Nature*, 544(7649), 2017, pp. 180–4.
8 Iain Davidson & William Noble, 'Why the First Colonisation of the Australian Region is the Earliest Evidence of Modern Human Behaviour', *Archaeology in Oceania*, 27(3), 1992, pp. 113–19; Jane

Balme, 'Of Boats and String: The Maritime Colonisation of Australia', *Quaternary International*, 285, 2013, pp. 68–75.

9 Jim Allen & James F O'Connell, 'Getting from Sunda to Sahul', in Geoffrey Clark, Foss Leach & Sue O'Connor (eds), *Islands of Inquiry: Colonization, Seafaring and the Archaeology of Maritime Landscapes*, Terra Australis 29 (ANU E Press, Canberra, 2008), pp. 31–46.

10 Sue O'Connor et al., 'Hominin Dispersal and Settlement East of Huxley's Line: The Role of Sea Level Changes, Island Size, and Subsistence Behavior', *Current Anthropology*, 58(S17), 2017, pp. S567–82. See also Jeffrey T Clark, 'Early Settlement in the Indo-Pacific', *Journal of Anthropological Archaeology*, 10(1), 1991, pp. 27–53.

11 Butlin, *Economics and the Dreamtime*, pp. 33–5.

12 Butlin, *Economics and the Dreamtime*, p. 8.

13 Peter Veth et al., 'Plants Before Farming: The Deep History of Plant-Use and Representation in the Rock Art of Australia's Kimberley Region', *Quaternary International*, 489, 2018, pp. 26–45; Damien Finch et al., '12,000-Year-Old Aboriginal Rock Art from the Kimberley Region, Western Australia', *Science Advances*, 6(6), 2020; Roger A Luebbers, 'Ancient Boomerangs Discovered in South Australia', *Nature*, 253, 1975, p. 39. For an overview of Ice Age bone technology in Australia, see Michelle C Langley, Sue O'Connor & Ken Aplin, 'A >46,000-Year-Old Kangaroo Bone Implement From Carpenter's Gap 1 (Kimberley, Northwest Australia)', *Quaternary Science Reviews*, 154, 2016, pp. 199–213.

14 Sonia Harmand et al., '3.3-Million-Year-Old Stone Tools from Lomekwi 3, West Turkana, Kenya', *Nature*, 521(7552), 2015, pp. 310–15; David R. Braun et al., 'Earliest Known Oldowan Artifacts at >2.58 Ma from Ledi-Geraru, Ethiopia, Highlight Early Technological Diversity', *Proceedings of the National Academy of Sciences*, 116(2), 2019, pp. 11712–17; Clarkson et al., 'Human Occupation of Northern Australia by 65,000 Years Ago'.

15 Jim M Bowler et al., 'Pleistocene Human Remains from Australia: A Living Site and Human Cremation from Lake Mungo, Western New South Wales', *World Archaeology*, 2(1), 1970, pp. 39–60.

16 Francis P Dickson, *Australian Stone Hatchets: A Study in Design and Dynamics* (Academic Press, Sydney, 1981); Anne Ford & Peter Hiscock, 'Axe Quarrying, Production, and Exchange in Australia and

New Guinea', in McNiven & David (eds), *The Oxford Handbook of the Archaeology of Indigenous Australia and New Guinea*.

17 Val Attenbrow & Nina Kononenko, 'Microscopic Revelations: The Forms and Multiple Uses of Ground-Edged Artefacts of the New South Wales Central Coast, Australia', *Technical Reports of the Australian Museum*, 29, 2019, pp. 1–100.

18 Attenbrow & Kononenko, 'Microscopic Revelations'.

19 Ford & Hiscock, 'Axe Quarrying, Production, and Exchange in Australia and New Guinea'.

20 Edward B Tylor, *Researches into the Early History of Mankind and the Development of Civilisation* (John Murray, London, 1865).

21 Ian J McNiven, 'Primordialising Aboriginal Australians: Colonialist Tropes and Eurocentric Views on Behavioural Markers of Modern Humans', in Martin Porr & Jacqueline Matthews (eds), *Interrogating Human Origins: Decolonisation and the Deep Human Past* (Routledge, Oxford & New York, 2020), pp. 96–111.

22 Grahame Clark, *World Prehistory: An Outline* (Cambridge University Press, Cambridge, 1961).

23 Carmel White [nee Schrire], 'Early Stone Axes in Arnhem Land', *Antiquity* 41(162), 1965, pp. 149–52; Carmel Schrire, *The Alligator Rivers: Prehistory and Ecology in Western Arnhem Land*, Terra Australis 7 (Department of Prehistory, Research School of Pacific Studies, Australian National University, Canberra, 1982).

24 Jean-Michel Geneste et al., 'The Origins of Ground-Edge Axes: New Findings from Nawarla Gabarnmang, Arnhem Land (Australia) and Global Implications for the Evolution of Fully Modern Humans', *Cambridge Archaeological Journal*, 22(1), 2012, pp. 1–17.

25 Tsutsumi Takashi, 'MIS3 Edge-Ground Axes and the Arrival of the First *Homo sapiens* in the Japanese Archipelago', *Quaternary International*, 248, 2012, pp. 70–8.

26 Clarkson et al., 'Human Occupation of Northern Australia by 65,000 Years Ago'.

27 Geneste et al., 'The Origins of Ground-Edge Axes'.

28 Ford & Hiscock, 'Axe Quarrying, Production, and Exchange in Australia and New Guinea'.

29 Ford & Hiscock, 'Axe Quarrying, Production, and Exchange in Australia and New Guinea'.
30 Jim Allen (ed.), *Report of the Southern Forests Archaeological Project, Volume 1: Site Descriptions, Stratigraphies and Chronologies* (School of Archaeology, La Trobe University, Bundoora, Vic., 1996).
31 Richard Cosgrove, 'Forty-Two Degrees South: The Archaeology of Late Pleistocene Tasmania', *Journal of World Prehistory*, 13(4), 1999, pp. 371–73.
32 Ian J McNiven, 'Technological Organisation and Settlement in SW Tasmania after the Glacial Maximum', *Antiquity*, 68(258), 1994, pp. 75–82.
33 Cosgrove, 'Forty-Two Degrees South', pp. 357–402; McNiven, 'Technological Organisation and Settlement in SW Tasmania after the Glacial Maximum'.
34 RF Fudali & RJ Ford, 'Darwin Glass and Darwin Crater: A Progress Report', *Meteoritics*, 14(3), 1979, pp. 283–96; John A Westgate et al., 'New Fission-Track Ages of Australasian Tektites Define Two Age Groups: Discriminating Between Formation and Reset Ages', *Quaternary Geochronology*, 66, 2021.
35 Rhys Jones, 'From Kakadu to Kutikina: The Southern Continent at 18,000 Years Ago', in Clive Gamble & Olga Soffer (eds), *The World at 18,000 BP, Volume 2: Low Latitudes* (Unwin Hyman, London, 1990), pp. 264–95. See also Cosgrove, 'Forty-Two Degrees South', p. 374; McNiven, 'Technological Organisation and Settlement in SW Tasmania after the Glacial Maximum', p. 78.
36 Sandra Bowdler, *Hunter Island, Hunter Hill: Archaeological Investigations of a Prehistoric Tasmanian Site*, Terra Australis 8 (Department of Prehistory, Research School of Pacific Studies, Australian National University, Canberra, 1984), p. 122; Richard Wright, 'Flaked Stone Material from GGW-1', *Memoirs of the National Museum of Victoria*, 30, 1970, pp. 79–92.
37 Ian J McNiven, 'Backed to the Pleistocene', *Archaeology in Oceania*, 35, 2000, pp. 48–52.
38 Patrick Schmidt et al., 'Heat Treatment in the South African Middle Stone Age: Temperature Induced Transformations of Silcrete and Their

Technological Implications', *Journal of Archaeological Science*, 40(9), 2013), pp. 3519–31.

39 Patrick Schmidt et al., 'When Was Silcrete Heat Treatment Invented in South Africa?', *Palgrave Communications*, 6(1), 2020, pp. 1–10.

40 Patrick Schmidt & Peter Hiscock, 'The Antiquity of Australian Silcrete Heat Treatment: Lake Mungo and the Willandra Lakes', *Journal of Human Evolution*, 142, 2020.

41 Harvey Johnston & Peter Clarke, 'Willandra Lakes Archaeological Investigations 1968–98', *Archaeology in Oceania*, 33(3), 1998, pp. 105–99.

42 Patrick Schmidt & Peter Hiscock, 'The Antiquity of Australian Silcrete Heat Treatment: Lake Mungo and the Willandra Lakes'.

43 Patrick Schmidt & Peter Hiscock, 'Early Silcrete Heat Treatment in Central Australia: Puritjarra and Kulpi Mara', *Archaeological and Anthropological Sciences*, 12(188), 2020, pp. 1–7.

44 Patrick Schmidt & Peter Hiscock, 'Evolution of Silcrete Heat Treatment in Australia: A Regional Pattern on the South-East Coast and Its Evolution Over the Last 25 ka', *Journal of Paleolithic Archaeology*, 2, 2019, pp. 74–97.

3. ICE AGE BURIALS AND FUNERALS

1 Ian McNiven, 'The Double Island Point Aboriginal Burials, Coastal Southeast Queensland', *Australian Archaeology*, 32(1), 1991, pp. 10–16.

2 Details of the excavations can be found in Alan Thorne's 1975 PhD thesis from the University of Sydney, titled 'Kow Swamp and Lake Mungo: Towards an Osteology of Early Man in Australia'. See also Alan G Thorne & Philip G Macumber, 'Discoveries of Late Pleistocene Man at Kow Swamp, Australia', *Nature*, 238, 1972, pp. 316–19. Thorne took all of the excavated materials from Kow Swamp to the Australian National University in Canberra for subsequent analysis – see Julie Lahn, *Finders Keepers, Losers Weepers: A 'Social History' of the Kow Swamp Remains*, Ngulaig 15 (Aboriginal and Torres Strait Islander Studies Unit, University of Queensland, St Lucia, 1996), p. 23.

3 Alan L West, 'Aboriginal Man at Kow Swamp, Northern Victoria: The Problem of Locating the Burial Site of the KS1 Skeleton', *The Artefact*,

2(2), 1977, pp. 19–30; Thorne & Macumber, 'Discoveries of Late Pleistocene Man at Kow Swamp, Australia', p. 316.

4 Thorne & Macumber, 'Discoveries of Late Pleistocene Man at Kow Swamp, Australia', p. 316.

5 Phillip G Macumber, 'The Geology and Palaeohydrology of the Kow Swamp Fossil Hominid Site, Victoria, Australia', *Journal of the Geological Society of Australia*, 24(5–6), 1977, pp. 307–20.

6 Tim Stone & Matthew L Cupper, 'Last Glacial Maximum Ages for Robust Humans at Kow Swamp, Southern Australia', *Journal of Human Evolution*, 45(2), 2003, pp. 99–111.

7 Thorne & Macumber, 'Discoveries of Late Pleistocene Man at Kow Swamp, Australia'.

8 Peter Brown, 'Artificial Cranial Deformation: A Component in the Variation in Pleistocene Australian Aboriginal Crania', *Archaeology in Oceania*, 16(3), 1981, pp. 156–67.

9 Josephine Flood, *Archaeology of the Dreamtime* (Collins, Sydney, 1989), p. 61.

10 Colin Pardoe, 'The Cemetery as Symbol: The Distribution of Prehistoric Aboriginal Burial Grounds in Southeastern Australia', *Archaeology in Oceania*, 23(1), 1988, pp. 1–16.

11 Colin Pardoe, 'The Demographic Basis of Human Evolution in Southeastern Australia', in Betty Meehan & Neville White (eds), *Hunter-Gatherer Demography*, Oceania Monograph 39 (Oceania Publications, Sydney, 1990), pp. 59–70.

12 The repatriation of the Kow Swamp burials to the Yorta Yorta (Eucha community) generated considerable debate among the Australian archaeological community – see Julie Lahn, *Finders Keepers, Losers Weepers*; D John Mulvaney, 'Past Regained, Future Lost: The Kow Swamp Pleistocene Burials', *Antiquity*, 65(246), 1991, pp. 12–21; Sandra Bowdler, 'Unquiet Slumbers: The Return of the Kow Swamp Burials', *Antiquity*, 66(250), 1992, pp. 103–6.

13 For example, see Billy Griffiths, *Deep Time Dreaming: Uncovering Ancient Australia* (Black Inc., Carlton, Vic., 2018); Andrew Pike & Ann McGrath (directors), *Message from Mungo* [documentary film],

2015; James M Bowler, *Lake Mungo: Window to Australia's Past* [CD-ROM] (School of Earth Sciences, University of Melbourne, Parkville, 2002); Helen Lawrence (ed.), *Mungo Over the Millennia: The Willandra Landscape and Its People* (Maygog Publishing, Sorell, Tas., 2006).

14 Griffiths, *Deep Time Dreaming*, p. 123.

15 Griffiths, *Deep Time Dreaming*, p. 124.

16 Bowler et al., 'Pleistocene Human Remains from Australia'; Griffiths, *Deep Time Dreaming*, p. 121.

17 Bowler et al., 'Pleistocene Human Remains from Australia'.

18 Griffiths, *Deep Time Dreaming*, p. 122.

19 Bowler et al., 'Pleistocene Human Remains from Australia', p. 56.

20 Steve G Webb, *The First Boat People* (Cambridge University Press, Cambridge, 2006), pp. 208, 233–4.

21 James M Bowler et al., 'New Ages for Human Occupation and Climatic Change at Lake Mungo, Australia', *Nature*, 421(6925), 2003, pp. 837–40.

22 Betty Hiatt [nee Meehan], 'Cremation in Aboriginal Australia', *Mankind*, 7(2), 1969, pp. 104–19.

23 Howard Williams, 'Towards an Archaeology of Cremation', in Christopher W Schmidt & Steven A Symes (eds), *The Analysis of Burnt Human Remains*, 2nd edn (Academic Press, Amsterdam, 2015), pp. 259–93; Joel D Irish, Ben A Potter & Joshua D Reuther, 'An 11,500-Year-Old Human Cremation from Eastern Beringia (Central Alaska)', in Schmidt & Symes (eds), *The Analysis of Burnt Human Remains*, pp. 295–306; Christopher W Schmidt et al., 'Early Archaic Cremations from Southern Indiana', in Schmidt & Symes (eds), *The Analysis of Burnt Human Remains*, pp. 227–37; Helen Lewis et al., 'Terminal Pleistocene to Mid-Holocene Occupation and an Early Cremation Burial at Ille Cave, Palawan, Philippines', *Antiquity*, 82(316), 2008, pp. 318–35; Ben Potter et al., 'A Terminal Pleistocene Child Cremation and Residential Structure from Eastern Beringia', *Science*, 331(6020), 2011, pp. 1058–62; Miguel Á Moreno-Ibáñez et al., 'Inhumation and Cremation: Identifying Funerary Practices and Reuse of Space Through Forensic Taphonomy at Cova Foradada (Calafell, Spain)', *Archaeological and Anthropological Sciences*, 14(4), 2022, pp. 57.

24 James M Bowler & Alan G Thorne, 'Human Remains from Lake Mungo: Discovery and Excavation of Lake Mungo III', in Kirk & Thorne (eds), *The Origin of the Australians*, pp. 127–38. The remains of a man designated Mungo II found near Mungo Lady comprised 'approximately thirty small fragments, mostly of the cranium and vertebra' and have attracted little research attention or publicity – see Bowler et al., 'Pleistocene Human Remains from Australia', p. 56 and Webb, *The First Boat People*, p. 209.
25 Bowler & Thorne, 'Human Remains from Lake Mungo', p. 129; Bowler et al., 'New Ages for Human Occupation and Climatic Change at Lake Mungo, Australia', p. 840.
26 Webb, *The First Boat People*, pp. 208, 234.
27 Stephen G Webb, *The Willandra Lakes Hominids* (Department of Prehistory, Research School of Pacific Studies, Australian National University, Canberra, 1989), pp. 52–7; Stephen Webb, *Palaeopathology of Aboriginal Australians: Health and Disease Across a Hunter-Gatherer Continent* (Cambridge University Press, Cambridge, 1995), pp. 42–9.
28 Webb, *The Willandra Lakes Hominids*, pp. 66, 79; Webb, Palaeopathology of Aboriginal Australians, pp. 59–61; Webb, *The First Boat People*, pp. 226–7.
29 Webb, *The Willandra Lakes Hominids*, p. 67; Webb, *The First Boat People*, p. 229.
30 Bowler & Thorne, 'Human Remains from Lake Mungo', p. 136.
31 Bowler et al., 'New Ages for Human Occupation and Climatic Change at Lake Mungo, Australia'.
32 Rainer Grün et al., 'U-series and ESR Analyses of Bones and Teeth Relating to the Human Burials from Skhul', *Journal of Human Evolution*, 49(3), 2005, pp. 316–34.
33 Yaroslav V Kuzmin, 'The Older, the Better? On the Radiocarbon Dating of Upper Palaeolithic Burials in Northern Eurasia and Beyond', *Antiquity*, 93(370), 2019, pp. 1061–71.
34 Paul B Pettitt, 'The Neanderthal Dead: Exploring Mortuary Variability in Middle Palaeolithic Eurasia', *Before Farming*, 1(4), 2002, pp. 1–19; William Rendu et al., 'Evidence Supporting an Intentional Neandertal

Burial at La Chapelle-aux-Saints', *Proceedings of the National Academy of Sciences*, 111(1), 2014, pp. 81–6; Harold L Dibble et al., 'A Critical Look at Evidence from La Chapelle-aux-Saints Supporting an Intentional Neandertal Burial', *Journal of Archaeological Science*, 53, 2015, pp. 649–57.

35 Francesco d'Errico & Lucinda Backwell, 'Earliest Evidence of Personal Ornaments Associated with Burial: The Conus Shells from Border Cave', *Journal of Human Evolution*, 93, 2016, pp. 91–108.

36 Pettitt, 'The Neanderthal Dead', pp. 9, 17; Paul Pettitt, *The Palaeolithic Origins of Human Burial* (Routledge, London & New York, 2011), Chapter 6.

37 Scott Cane, *First Footprints: The Epic Story of the First Australians* (Allen & Unwin, Sydney, 2013), p. 95.

38 Harvey Johnston & Peter Clarke, 'Willandra Lakes Archaeological Investigations 1968–98', *Archaeology in Oceania*, 33(3), 1998, pp. 105–99; Webb, *The Willandra Lakes Hominids*, p. 13; Webb, *The First Boat People*, p. 225.

39 Webb, *The First Boat People*, pp. 221–2.

40 Webb, *The First Boat People*, p. 226.

41 Cane, *First Footprints*, pp. 93, 95.

42 Griffiths, *Deep Time Dreaming*, pp. 133–40; Lawrence (ed.), *Mungo Over the Millennia*, 2006. Mungo Lady was repatriated to Traditional Owners in 1991 and Mungo Man in 2017. The issue of whether to rebury the Lake Mungo ancestral remains directly into the ground or preserve them in an underground keeping place continues to be discussed among Traditional Owners.

43 Johnston & Clarke, 'Willandra Lakes Archaeological Investigations 1968–98', p. 117.

4. ENHANCED NETWORKING, TRADING AND SHARING

1 Margo Neale & Lynne Kelly, *Songlines: The Power and Promise* (Thames & Hudson, Port Melbourne, 2020). For a discussion of links between trade routes and Songlines, see Franz J Micha, 'Trade and Change in Australian Aboriginal Cultures', pp. 285–313.

2 D John Mulvaney, 'The Chain of Connection: The Material Evidence', in Nicolas Peterson (ed.), *Tribes and Boundaries in Australia*, Social Anthropology Series No. 10 (Australian Institute of Aboriginal Studies, Canberra, 1976), p. 80; Ian Keen, *Aboriginal Economy & Society: Australia at the Threshold of Colonisation* (Oxford University Press, South Melbourne, 2004), pp. 354, 368. The major exception to trade in a consumable is the large-scale trade in pituri, a psychoactive plant-based narcotic drug that was traded over distances of 900 kilometres and across a 500,000-square-kilometre area of eastern Central Australia. Much pituri was obtained during expeditions of men travelling hundreds of kilometres. No archaeological insights are available on the antiquity of pituri use or trade. See Pamela Watson, *This Precious Foliage: A Study of the Aboriginal Psycho-Active Drug Pituri*, Oceania Monograph 26 (University of Sydney, Sydney, 1983); Cane, *First Footprints*, p. 231; Mike Smith, *The Archaeology of Australia's Deserts* (Cambridge University Press, Cambridge, 2013), pp. 292–4. Some historical (archival and oral) and archaeological evidence supports that the Gunditjmara of south-west Victoria smoked eels for export – see Heather Builth, *Ancient Aboriginal Aquaculture Rediscovered: The Archaeology of an Australian Cultural Landscape* (Lambert Academic, Saarbrucken, Germany, 2014).

3 Alfred C Haddon, 'Trade', in Alfred C Haddon, *Reports of the Cambridge Anthropological Expedition to Torres Straits, Volume 5: Sociology, Magic and Religion of the Western Islanders* (Cambridge University Press, Cambridge, 1904), pp. 293–7.

4 David Lawrence, 'Customary Exchange Across Torres Strait', *Memoirs of the Queensland Museum*, 34(2), 1994, pp. 214–446.

5 Ian J McNiven, 'Enmity and Amity: Reconsidering Stone-Headed Club (Gabagaba) Procurement and Trade in Torres Strait', *Oceania*, 69, 1998, pp. 94–115.

6 Lawrence, 'Customary Exchange Across Torres Strait'.

7 Johannes Falkenberg, *Kin and Totem: Group Relations of Australian Aborigines in the Port Keats District* (Oslo University Press, Oslo, 1962), pp. 152–7.

8 Daisy Bates, 'Great Aboriginal Trade Route', *The Australasian*, 1 November 1930, p. 4.

9 Corey JA Bradshaw et al., 'Minimum Founding Populations for the First Peopling of Sahul'.

10 Ian Keen, *Aboriginal Economy & Society*, p. 125.

11 Susan O'Connor, Peter Veth & Nicola Hubbard, 'Changing Interpretations of Postglacial Human Subsistence and Demography in Sahul', in Mike A Smith, Matthew Spriggs & Barry Fankhauser (eds), *Sahul in Review: Pleistocene Archaeology in Australia, New Guinea and Island Melanesia*, Occasional Papers in Prehistory 24 (Department of Prehistory, Research School of Pacific Studies, Australian National University, Canberra, 1993), pp. 95–105.

12 Joseph Birdsell's proposed rapid population model for Australia is theoretically feasible, according to Corey Bradshaw and colleagues: Birdsell, in Allen, Golson & Jones (eds), *Sunda and Sahul*, pp. 113–67; Bradshaw et al., 'Stochastic Models Support Rapid Peopling of Late Pleistocene Sahul'. Achievement of historically known population levels only within the past few thousand years has been argued variously by John Beaton, Harry Lourandos, and Alan Williams and colleagues: John M Beaton, 'Comment: Does Intensification Account for Changes in the Australian Holocene Archaeological Record?', *Archaeology in Oceania*, 8, 1983, pp. 94–7; Harry Lourandos, 'Pleistocene Australia: Peopling a Continent' in Olga Soffer (ed.), *The Pleistocene Old World: Regional Perspectives* (Plenum Press, New York, 1987), pp. 147–65; Alan N Williams et al., 'A Continental Narrative: Human Settlement Patterns and Australian Climate Change Over the Last 35,000 Years', *Quaternary Science Reviews*, 123, 2015, pp. 91–112; Alan N Williams et al., 'Holocene Demographic Changes and the Emergence of Complex Societies in Prehistoric Australia', *PLoS One*, 10(6), 2015.

13 Jo McDonald, 'Archaic Faces to Headdresses: The Changing Role of Rock Art Across the Arid Zone', in Peter Veth, Mike Smith & Peter Hiscock (eds), *Desert Peoples: Archaeological Perspectives* (Blackwell, Malden, MA, 2005), pp. 116–41.

14 Ken Mulvaney, 'About Time: Toward a Sequencing of the Dampier Archipelago Petroglyphs of the Pilbara Region, Western Australia', *Records of the Western Australian Museum*, 79, 2011, pp. 30–49; Jo McDonald, 'I Must Go Down to the Seas Again: Or, What Happens

When the Sea Comes to You? Murujuga Rock Art as an Environmental Indicator for Australia's North-West', *Quaternary International*, 385, 2015, pp. 124–35.

15 McDonald, 'Archaic Faces to Headdresses', p. 136.

16 Clarkson et al., 'Human Occupation of Northern Australia by 65,000 Years Ago'; Franklin & Habgood, 'Modern Human Behaviour and Pleistocene Sahul in Review'.

17 Jillian Huntley, 'Australian Indigenous Ochres: Use, Sourcing, and Exchange', in McNiven & David (eds), *The Oxford Handbook of the Archaeology of Indigenous Australia and New Guinea*; Ian J McNiven & Bruno David, 'Rock-Art of Torres Strait: Overview of Sites and Ochre Sources', *Memoirs of the Queensland Museum: Cultural Heritage Series*, 3(1), 2004, pp. 209–25.

18 Tammy Hodgskiss, 'Ochre Use in the Middle Stone Age', in Mark Aldenderfer (ed.), *The Oxford Research Encyclopedia of Anthropology* (online) (Oxford University Press, Oxford, 2020); Christopher S Henshilwood et al., 'A 100,000-Year-Old Ochre-Processing Workshop at Blombos Cave, South Africa', *Science*, 334, 2011, pp. 219–22; Ian Watts, Michael Chazan & Jayne Wilkins, 'Early Evidence for Brilliant Ritualized Display: Specularite Use in the Northern Cape (South Africa) Between ~500 and ~300 ka', *Current Anthropology*, 57(3), 2016, pp. 287–310; Jan M Burdukiewicz, 'The Origin of Symbolic Behavior of Middle Palaeolithic Humans: Recent Controversies', *Quaternary International*, 326, 2014, pp. 398–405; Antonio G Sagona (ed.), *Bruising the Red Earth: Ochre Mining and Ritual in Aboriginal Tasmania* (Melbourne University Press, Carlton, Vic., 1994).

19 Harvey Johnston & Peter Clarke, 'Willandra Lakes Archaeological Investigations 1968–98', *Archaeology in Oceania*, 33(3), 1998, pp. 105–99. Ochre-sourcing expert Jillian Huntley makes the point that comparing the chemical fingerprints of Mungo Man's red ochre with regional ochre outcrops would help identify exactly where the Old People of Lake Mungo obtained their ochre – see Huntley, 'Australian Indigenous Ochres'.

20 Nicolas Peterson & Ronald Lampert, 'A Central Australian Ochre Mine', *Records of the Australian Museum*, 37, 1985, pp. 1–9; Cane, *First Footprints*, p. 232.

21 Peterson & Lampert, 'A Central Australian Ochre Mine', pp. 6–7.
22 Lampert, 'A Central Australian Ochre Mine', p. 1.
23 Lampert, 'A Central Australian Ochre Mine', p. 2.
24 Smith, *The Archaeology of Australia's Deserts*, pp. 280–1. The Puritjarra rock-shelter ochre fragments were matched to the Karrku ochre mine by chemical fingerprinting – see Mike A Smith, Barry Fankhauser & Meliita Jercher, 'The Changing Provenance of Red Ochre at Puritjarra Rock Shelter, Central Australia: Late Pleistocene to Present', *Proceedings of the Prehistoric Society*, 64, 1998, pp. 275–92.
25 Smith, *The Archaeology of Australia's Deserts*, pp. 281, 296.
26 Kate Morse, 'New Radiocarbon Dates from North West Cape, Western Australia: A Preliminary Report', in Smith, Spriggs & Fankhauser (eds), *Sahul in Review*, pp. 155–63.
27 Jane Balme & Kate Morse, 'Shell Beads and Social Behaviour in Pleistocene Australia', *Antiquity*, 80(310), 2006, pp. 799–811.
28 Jane Balme, Sue O'Connor & Michelle Langley, 'Marine Shell Ornaments in North Western Australian Archaeological Sites: Different Meanings Over Time and Space', in Michelle Langley et al. (eds), *The Archaeology of Portable Art: Southeast Asian, Pacific and Australian Perspectives* (Routledge, London, 2018), pp. 258–73; Balme & Morse, 'Shell Beads and Social Behaviour in Pleistocene Australia', p. 807.
29 Balme & Morse, 'Shell Beads and Social Behaviour in Pleistocene Australia', p. 809.
30 Sue O'Connor, *30,000 Years of Aboriginal Occupation: Kimberley, North West Australia*, Terra Australis 14 (Research School of Pacific and Asian Studies, Australian National University, Canberra, 1999), p. 121.
31 Harry Lourandos, *Continent of Hunter-Gatherers: New Perspectives in Australian Prehistory* (Cambridge University Press, Cambridge, 1997), pp. 299–300.
32 Williams et al., 'A Continental Narrative'; Williams et al., 'Holocene Demographic Changes and the Emergence of Complex Societies in Prehistoric Australia'.
33 Joseph B Birdsell, 'Realities and Transformations: The Tribes of the Western Desert of Australia', in Peterson (ed.), *Tribes and Boundaries*

in Australia, pp. 95–132; Lourandos, *Continent of Hunter-Gatherers*, pp. 38–43.

34 Harry Lourandos, 'Intensification and Australian Prehistory', in T Douglas Price & JA Brown (eds), *Prehistoric Hunter-Gatherers: The Emergence of Cultural Complexity* (Academic Press, Orlando, FL, 1985), pp. 385–423; Smith, *The Archaeology of Australia's Deserts*, p. 298.

35 Peter Hiscock, 'Technological Responses to Risk in Holocene Australia', *Journal of World Prehistory*, 8, 1994, pp. 267–92. Peter Veth, Peter Hiscock & Alan Williams, 'Are Tulas and ENSO Linked in Australia?', *Australian Archaeology*, 72(1), 2011, pp. 7–14; Peter Veth, 'Cycles of Aridity and Human Mobility: Risk Minimization Amongst Late Pleistocene Foragers of the Western Desert, Australia', in Veth, Smith & Hiscock (eds), *Desert Peoples*, pp. 100–15; Ian J McNiven, 'Precarious Islands: Kulkalgal Reef Island Settlement and High Mobility Across 700 km of Seascape, Central Torres Strait and Northern Great Barrier Reef', *Quaternary International*, 385, 2015, pp. 39–55.

36 For a list of the number of people attending various gatherings around Australia, see D John Mulvaney, 'The Chain of Connection', pp. 84–6.

37 Harry Lourandos, 'Palaeopolitics: Resource Intensification in Aboriginal Australia and Papua New Guinea', in Tim Ingold, David Riches & James Woodburn (eds), *Hunters and Gatherers 1: History, Evolution and Social Change* (Berg, New York/Oxford, 1991), pp. 148–60; Ian J McNiven & Ariana BJ Lambrides, 'Stone-Walled Fish Traps of Australia and New Guinea as Expressions of Enhanced Sociality', in McNiven & David (eds), *The Oxford Handbook of the Archaeology of Indigenous Australia and New Guinea*; Josephine Flood, *The Moth Hunters: Aboriginal Prehistory of the Australian Alps* (Australian Institute of Aboriginal Studies, Canberra, 1980).

38 Micha, 'Trade and Change in Australian Aboriginal Cultures'; Mulvaney, 'The Chain of Connection'.

39 The archaeological term 'tula' dates to the 1930s and is based on 'Koondi tulha', the ethnographically recorded term used by the Wongkanguru people of the Lake Eyre region for a hafted stone adze – see Ian J McNiven, 'Tula Adzes and Bifacial Points on the East Coast of Australia', *Australian Archaeology*, 36, 1993, p. 23. See also Mark W

Moore, 'The Tula Adze: Manufacture and Purpose', *Antiquity*, 78(299), 2004, pp. 61–73; Trudy Doelman & Grant WG Cochrane, 'Design Theory and the Australian Tula Adze', *Asian Perspectives*, 51(2), 2012, pp. 251–77; Tim R Maloney & India E Dilkes-Hall, 'Assessing the Spread and Uptake of Tula Adze Technology in the Late Holocene Across the Southern Kimberley of Western Australia', *Australian Archaeology*, 86(3), 2020, pp. 264–83; Veth, Hiscock & Williams, 'Are Tulas and ENSO Linked in Australia?'.

40 McNiven, 'Tula Adzes and Bifacial Points on the East Coast of Australia', pp. 22–33; Maloney & Dilkes-Hall, 'Assessing the Spread and Uptake of Tula Adze Technology in the Late Holocene Across the Southern Kimberley of Western Australia'.

41 Doelman & Cochrane, 'Design Theory and the Australian Tula Adze', pp. 255, 264; Peter Hiscock, 'A Cache of Tulas from the Boulia District, Western Queensland', *Archaeology in Oceania*, 23(2), 1988, pp. 60–70.

42 Ian J McNiven, 'Backed Artefacts as a Material Dimension of Social Inclusiveness', *Australian Archaeology*, 72, 2011, pp. 71–2.

43 Gail Robertson, Val Attenbrow & Peter Hiscock, 'Multiple Uses for Australian Backed Artefacts', *Antiquity*, 83(320), 2009, pp. 296–398.

44 Peter Hiscock, *Archaeology of Ancient Australia* (Routledge, London & New York, 2008), p. 156.

45 Sean Ulm & Ian J McNiven, 'Excavating Jiigurru', in Bronwyn Mitchell & Ruth Ridgway (eds), *Connections Across the Coral Sea: A Story of Movement* (Queensland Museum, South Brisbane, 2021), pp. 30–3.

46 Ian J McNiven, 'Coral Sea Cultural Interaction Sphere', in McNiven & David (eds), *The Oxford Handbook of the Archaeology of Indigenous Australia and New Guinea*; Ian J McNiven, 'Beyond Bridge and Barrier: Reconceptualising Torres Strait as a Co-Constructed Border Zone in Ethnographic Object Distributions Between Queensland and New Guinea', *Queensland Archaeological Research*, 25, 2022, pp. 25–46.

47 Fred D McCarthy, '"Trade" in Aboriginal Australia, and "Trade" Relationships with Torres Strait, New Guinea and Malaya', *Oceania*, 9(4), 1939, pp. 405–39; 10(1), 1940, pp. 80–104; 10(2), 1940, pp. 171–95.

48 Vicky Winton et al., 'Thuwarri Thaa (Wilgie Mia): Ancient Aboriginal Mining Technology in Cultural Context', in Cat Kutay et al. (eds),

Indigenous Engineering for an Enduring Culture (Cambridge Scholars Publishing, Newcastle upon Tyne, 2022), pp. 405–15.

49 Smith, *The Archaeology of Australia's Deserts*, p. 278.

50 Smith, *The Archaeology of Australia's Deserts*, p. 279.

51 Smith, *The Archaeology of Australia's Deserts*, p. 280; Henry P Woodward, *A Geological Reconnaissance of a Portion of the Murchison Goldfield*, Western Australia Geological Survey Bulletin 57 (Fred WM Simpson Government Printer, Perth, 1914), p. 83.

52 Smith, *The Archaeology of Australia's Deserts*, pp. 276–8. For a detailed and remarkable synthesis of information about the Pukardu ochre mine, see Philip Jones, *Ochre and Rust: Artefacts and Encounters on Australian Frontiers* (Wakefield Press, Kent Town, SA, 2007), Chapter 9.

53 Jones, *Ochre and Rust*, pp. 351–2; Isabel McBryde, 'Travellers in Storied Landscapes: A Case Study in Exchanges and Heritage', *Aboriginal History*, 24, 2000, pp. 152–74.

54 Alfred W Howitt, *The Native Tribes of South-East Australia* (Macmillan, London, 1904), pp. 711–12.

55 Samuel Gason, 'The Manners and Customs of the Dieyerie Tribe of Australian Aborigines', in JD Woods (ed.), *The Native Tribes of South Australia* (ES Wigg & Son, Adelaide, 1879), pp. 253–307.

56 Jones, *Ochre and Rust*, p. 350.

57 TA Masey, 'The Red Ochre Caves of the Blacks', *Port Augusta Dispatch and Flinders' Advertiser*, 9 June 1882, p. 3; Charles Mountford cited in McCarthy, '"Trade" in Aboriginal Australia, and "Trade" Relationships with Torres Strait, New Guinea and Malaya', *Oceania*, 9(4), 10(1), 1940, p. 88; Jones, *Ochre and Rust*, pp. 353–4.

58 Peterson & Lampert, 'A Central Australian Ochre Mine', p. 1.

59 Jones, *Ochre and Rust*, pp. 355–62. See also McBryde, 'Travellers in Storied Landscapes'.

60 A Peter Elkin, 'Cult Totemism and Mythology in Northern South Australia', *Oceania*, 5(2), 1934, pp. 171–92.

61 Smith, *The Archaeology of Australia's Deserts*, p. 277.

62 McCarthy, '"Trade" in Aboriginal Australia, and "Trade" Relationships with Torres Strait, New Guinea and Malaya', *Oceania*, 9(4), 10(1), 1940, pp. 92–5; Charles P Mountford & Alison Harvey, 'A Survey of

Australian Aboriginal Pearl and Baler Shell Ornaments', *Records of the South Australian Museum*, 6(2), 1938, pp. 115–35; Kim Akerman, 'Aboriginal Baler Shell Objects in Western Australia', *Mankind*, 9, 1973, pp. 124–5; Mulvaney, 'The Chain of Connection', p. 84.

63 Mountford & Harvey, 'A Survey of Australian Aboriginal Pearl and Baler Shell Ornaments'; McCarthy, '"Trade" in Aboriginal Australia, and "Trade" Relationships with Torres Strait, New Guinea and Malaya', *Oceania*, 9(4), 10(1), 1940, pp. 92–5; Mulvaney, 'The Chain of Connection'; Isabel McBryde, 'Goods from Another Country: Exchange Networks and the People of the Lake Eyre Basin', in Mulvaney & White (eds), *Australians to 1788*, pp. 253–73.

64 Lawrence, 'Customary Exchange Across Torres Strait'.

65 Mike A Smith & Peter M Veth, 'Radiocarbon Dates for Shell in the Great Sandy Desert', *Australian Archaeology*, 58, 2004, pp. 37–8.

66 McBryde, 'Goods from Another Country'; Smith, *The Archaeology of Australia's Deserts*, pp. 282–7.

67 Isabel McBryde, 'The Landscape Is a Series of Stories: Grindstones, Quarries and Exchange in Aboriginal Australia: A Lake Eyre Case Study', in A Ramos-Millan & A Bustillo (eds), *Siliceous Rocks and Culture* (University of Granada Press, Granada, 1997), pp. 587–607; Smith, *The Archaeology of Australia's Deserts*, pp. 283–4.

68 McBryde, 'Goods from Another Country'; McBryde, 'The Landscape Is a Series of Stories'.

69 Veth, 'Cycles of Aridity and Human Mobility', p. 110.

70 Anne Ford & Peter Hiscock, 'Axe Quarrying, Production, and Exchange in Australia and New Guinea', in McNiven & David (eds), *The Oxford Handbook of the Archaeology of Indigenous Australia and New Guinea*.

71 Peter Hiscock, 'Standardised Axe Manufacture at Mount Isa', in Ingereth Macfarlane, Mary-Jane Mountain & Robert Paton (eds), *Many Exchanges: Archaeology, History, Community and the Work of Isabel McBryde*, Aboriginal History Monograph 11 (Aboriginal History Inc., Canberra, 2005), pp. 287–300.

72 Smith, *The Archaeology of Australia's Deserts*, p. 290.

73 McBryde, 'Goods from Another Country'; McBryde, 'The Landscape Is a Series of Stories'; Iain Davidson et al., 'Archaeology in Another

Country: Exchange and Symbols in North-West Central Queensland', in Macfarlane, Mountain & Paton (eds), *Many Exchanges*, pp. 103–30.

74 Wurundjeri are one of the five communities that comprise the Kulin nation of south-central Victoria. They and their neighbours the Boonwurrung, Taungurung and Dja Dja Wurrung make up the eastern Kulin, while the Wathaurung are the western Kulin.

75 James Bonwick, *Port Phillip Settlement* (Sampson Low, London, 1883).

76 William Blandowski, 'Personal Observations Made in an Excursion Towards the Central Parts of Victoria, including Mount Macedon, McIvor, and Black Ranges', *Transactions of the Philosophical Society of Victoria*, 1, 1855, pp. 50–74.

77 William Barak's (Billibellary's nephew) description to anthropologist Alfred Howitt c. 1884, cited in Meyer Eidelson, *Melbourne Dreaming: A Guide to Important Places of the Past and Present* (Aboriginal Studies Press, Melbourne, 1997), p. 84.

78 Isabel McBryde & Alan Watchman, 'The Distribution of Greenstone Axes in Southeastern Australia', *Mankind*, 10(3), 1976, pp. 163–74; Isabel McBryde, 'Wil-im-ee Moor-ring: Or, Where Do Axes Come From?', *Mankind*, 11(3), 1978, pp. 354–82; Isabel McBryde, 'Exchange in South Eastern Australia: An Ethnohistorical Perspective', *Aboriginal History*, 8, 1984, pp. 132–53; Isabel McBryde, 'Kulin Greenstone Quarries: The Social Contexts of Production and Distribution for the Mt William Site', *World Archaeology*, 16, 1984, pp. 267–85; Robert Paton, 'Trading Places: Changing Social Values of the Mt William Aboriginal Stone Quarry', in Macfarlane, Mountain & Paton (eds), *Many Exchanges*, pp. 271–86.

79 Gunaikurnai Land and Waters Aboriginal Corporation, 'Our Story', Gunaikurnai Land and Waters Aboriginal Corporation website, 2023.

80 Isabel M McBryde & G Harrison, 'Valued Good or Valuable Stone? Consideration of Some Aspects of the Distribution of Greenstone Artefacts in South-Eastern Australia', in Foss Leach & Janet Davidson (eds), *Archaeological Studies of Pacific Stone Resources*, British Archaeological Reports International Series 104 (BAR, Oxford, 1981), pp. 183–209; Adam Brumm, '"The Falling Sky": Symbolic and Cosmological

Associations of the Mt William Greenstone Axe Quarry, Central Victoria, Australia', *Cambridge Archaeological Journal*, 20(2), 2010, pp. 179–96.

81 McNiven, 'Enmity and Amity'; Robert Paton, 'Speaking Through Stones: A Study from Northern Australia', *World Archaeology*, 26(2), 1994, pp. 172–84.

5. ENHANCED PLANTS AND ANIMALS

1 Michael J Rowland, 'Return of the "Noble Savage": Misrepresenting the Past, Present and Future', *Australian Aboriginal Studies*, 2, 2004, pp. 2–14.

2 Rhys Jones, 'Fire-stick Farming', *Australian Natural History*, 16, 1969, pp. 224–8.

3 Bill Gammage, *The Biggest Estate on Earth: How Aborigines Made Australia* (Allen & Unwin, Sydney, 2011); Bruce Pascoe, *Dark Emu: Black Seeds: Agriculture or Accident?* (Magabala Books, Broome, WA, 2014); Michael-Shawn Fletcher et al., 'Indigenous Knowledge and the Shackles of Wilderness', *Proceedings of the National Academy of Sciences*, 118(40), 2021; Michael-Shawn Fletcher, Tegan Hall & Andreas Nicholas Alexandra, 'The Loss of an Indigenous Constructed Landscape Following British Invasion of Australia: An Insight into the Deep Human Imprint on the Australian Landscape', *Ambio*, 50, 2021, pp. 138–49.

4 For a detailed synthesis and overview of Aboriginal and Torres Strait Islander practices of increasing the productivity of selected plants and animals in both terrestrial and freshwater and marine settings, see Ian J McNiven, Tiina Manne & Annie Ross, 'Enhanced Ecologies and Ecosystem Engineering: Strategies Developed by Aboriginal Australians to Increase the Abundance of Animal Resources', in McNiven & David (eds), *The Oxford Handbook of the Archaeology of Indigenous Australia and New Guinea*.

5 Peter Sutton & Keryn Walshe, *Farmers or Hunter-Gatherers? The Dark Emu Debate* (Melbourne University Press, Carlton, Vic., 2021), Chapter 2.

6 One of the first publications detailing a complex seasonal calendar for an Australian Aboriginal group was by anthropologist Donald Thomson on the Wik-Munkan people of western Cape York Peninsula in North

Queensland: Donald F Thomson, 'The Seasonal Factor in Human Culture Illustrated from the Life of a Contemporary Nomadic Group', *Proceedings of the Prehistoric Society*, 5(2), 1939, pp. 209–21.

7 Bureau of Meteorology, 'About the IWK Website', Indigenous Weather Knowledge website, 2016; CSIRO, 'Indigenous Seasonal Calendars', CSIRO website, n.d.

8 Suzanne M Prober, Michael H O'Connor & Fiona J Walsh, 'Australian Aboriginal Peoples' Seasonal Knowledge: A Potential Basis for Shared Understanding in Environmental Management', *Ecology and Society*, 16(2), 2011.

9 Emma Woodward et al., 'Utilising Indigenous Seasonal Knowledge to Understand Aquatic Resource Use and Inform Water Resource Management in Northern Australia', *Ecological Management and Restoration*, 13, 2012, pp. 58–64.

10 Emma Woodward & First Class in Graphic Design, *Wagiman Plants and Animals: Daly River, Northern Territory, Australia* [poster], CSIRO, 2012.

11 Damien A Fordham, Arthur Georges & Barry W Brook, 'Indigenous Harvest, Exotic Pig Predation and Local Persistence of a Long-Lived Vertebrate: Managing a Tropical Freshwater Turtle for Sustainability and Conservation', *Journal of Applied Ecology*, 45(1), 2008, pp. 52–62.

12 Bill Gammage & Bruce Pascoe, *Country: Future Fire, Future Farming* (Thames & Hudson, Port Melbourne, 2021).

13 Zena Cumpston, Michael-Shawn Fletcher & Lesley Head, *Plants: Past, Present and Future* (Thames & Hudson, Port Melbourne, 2022).

14 Beth Gott, 'Murnong – *Microseris scapigera*: A Study of a Staple Food of Victorian Aborigines', *Australian Aboriginal Studies*, 2, 1983, pp. 2–18.

15 John Morgan, *The Life and Adventures of William Buckley* (R Schicht, Sydney, 1996).

16 Thomas L Mitchell, *Three Expeditions into the Interior of Eastern Australia* (T & W Boone, London, 1839), pp. 90, 98.

17 Ian Clark (ed.), *The Journals of George Augustus Robinson, Volume 1* (Heritage Matters, Melbourne, 1998), pp. 162–3, 196–7, 308.

18 George Grey, *Journals of Two Expeditions of Discovery, Volume 2* (T & W Boone, London, 1841), pp. 12–28.

19 Brit Asmussen, 'Another Burning Question: Hunter-Gatherer Exploitation of *Macrozamia* spp.', *Archaeology in Oceania*, 44, 2009, pp. 42–9; Brit Asmussen, 'In a Nut Shell: The Identification and Archaeological Application of Experimentally Defined Correlates of *Macrozamia* Seed Processing', *Journal of Archaeological Science*, 37, 2010, pp. 2117–25; Wendy Beck, Richard Fullagar & Neville White, 'Archaeology from Ethnography: The Aboriginal Use of Cycad as an Example', in Betty Meehan & Rhys Jones (eds), *Archaeology with Ethnography: An Australian Perspective* (Department of Prehistory, Research School of Pacific Studies, Australian National University, Canberra, 1988), pp.137–47; Moya Smith, 'Late Pleistocene Zamia Exploitation in Southern Western Australia', *Archaeology in Oceania*, 17, 1982, pp. 117–21.

20 Carl Lumholtz, *Among Cannibals* (Scribner's Sons, New York, 1908); John M Beaton, 'Fire and Water: Aspects of Australian Aboriginal Management of Cycads', *Archaeology in Oceania*, 17(1), 1982, pp. 51–8.

21 John J Bradley, 'The Social, Economic and Historical Construction of Cycad Palms Among the Yanyuwa', in David, Barker & McNiven (eds), *The Social Archaeology of Australian Indigenous Societies*, pp. 161–81.

22 Beaton, 'Fire and Water'.

23 Bradley, 'The Social, Economic and Historical Construction of Cycad Palms Among the Yanyuwa', p. 180.

24 Bradley, 'The Social, Economic and Historical Construction of Cycad Palms Among the Yanyuwa', p. 180.

25 Åsa Ferrier & Richard Cosgrove, 'Aboriginal Exploitation of Toxic Nuts as a Late-Holocene Subsistence Strategy in Australia's Tropical Rainforests', in Simon G Haberle & Bruno David (eds), *Peopled Landscapes. Archaeological and Biogeographic Approaches to Landscape* (ANU E Press, Canberra, 2012), pp. 103–20.

26 Ferrier & Cosgrove, 'Aboriginal Exploitation of Toxic Nuts as a Late-Holocene Subsistence Strategy in Australia's Tropical Rainforests'.

27 Brit Asmussen, '"There Is Likewise a Nut . . .": A Comparative Ethnobotany of Aboriginal Processing Methods and Consumption of

Australian Bowenia, Cycas, Lepidozamia and Macrozamia Species', *Technical Reports of the Australian Museum*, 23(10), 2011, pp. 147–63.

28 Helene Marsh, Thomas J O'Shea & John E Reynolds III, *Ecology and Conservation of the Sirenia: Dugongs and Manatees* (Cambridge University Press, Cambridge, 2011), Chapter 8.

29 Joe Crouch et al., 'Berberass: Marine Resource Specialisation and Environmental Change in Torres Strait Over the Past 4000 Years', *Archaeology in Oceania*, 42, 2007, pp. 49–64.

30 Margaret Lawrie, *Myths and Legends of Torres Strait* (University of Queensland Press, St Lucia, 1970), pp. 57–60; Brian Robinson & Theo Tremblay (eds), *Ailan Currents: Contemporary Printmaking from the Torres Strait* (KickArts Contemporary Arts, Cairns, Qld, 2007).

31 For an overview of these rituals, see Ian J McNiven & Ricky Feldman, 'Ritually Orchestrated Seascapes: Hunting Magic and Dugong Bone Mounds in Torres Strait, NE Australia', *Cambridge Archaeological Journal*, 13(2), 2003, pp. 169–94; Ian J McNiven, 'Dugongs and Turtles as Kin: Relational Ontologies and Archaeological Perspectives on Ritualised Hunting by Coastal Indigenous Australians', in McNiven & David (eds), *The Oxford Handbook of the Archaeology of Indigenous Australia and New Guinea*.

32 Ian J McNiven, 'Between the Living and the Dead: Relational Ontologies and the Ritual Dimensions of Dugong Hunting Across Torres Strait', in Christopher Watts (ed.), *Archaeologies of Relationality: Humans, Animals, Things* (Routledge, London & New York, 2013), pp. 97–116.

33 McNiven & Feldman, 'Ritually Orchestrated Seascapes'; Ian J McNiven & Alice C Bedingfield, 'Past and Present Marine Mammal Hunting Rates and Abundances: Dugong (*Dugong dugon*) Evidence from Dabangai Bone Mound, Torres Strait', *Journal of Archaeological Science*, 35, 2008, pp. 505–15.

34 McNiven & Feldman, 'Ritually Orchestrated Seascapes'; Robert Skelly et al., 'The Ritual Dugong Bone Mounds of Koey Ngurtai, Torres Strait, Australia: Investigating Their Construction', *International Journal of Osteoarchaeology*, 21, 2011, pp. 32–54.

35 Ian J McNiven, 'Navigating the Human-Animal Divide: Marine Mammal Hunters and Rituals of Sensory Allurement', *World Archaeology*, 42(2), 2010, pp. 215–30.
36 McNiven, 'Dugongs and Turtles as Kin'.
37 Alfred C Haddon, 'Magic and Religion', in Haddon, *Reports of the Cambridge Anthropological Expedition to Torres Straits, Volume 5*, pp. 334–6, Plate XXI.2.
38 John J Bradley, 'Li-anthawirriyarra, People of the Sea: Yanyuwa Relations with Their Maritime Environment', PhD thesis, Northern Territory University, 1997, pp. 207–8, 321.
39 Cited in David R Moore (ed.), *Islanders and Aborigines at Cape York: An Ethnographic Reconstruction Based on the 1848–1850 'Rattlesnake' Journals of OW Brierly and Information He Obtained from Barbara Thompson* (Australian Institute of Aboriginal Studies, Canberra, 1979), pp. 168, 226.
40 Alfred C Haddon, *Reports of the Cambridge Anthropological Expedition to Torres Straits, Volume 1: General Ethnography* (Cambridge University Press, Cambridge, 1935), p. 87.
41 For a summary of published information on Torres Strait turtle-hunting rituals and associated sites and objects, see McNiven, 'Dugongs and Turtles as Kin'.
42 Haddon, *Reports of the Cambridge Anthropological Expedition to Torres Straits, Volume 1*, p. 69.
43 David R Harris & Barbara Ghaleb-Kirby, 'Mabuyag (Torres Strait) in the Mid-1980s: Archaeological Reconnaissance of the Island and Midden Excavations at Goemu', *Memoirs of the Queensland Museum: Culture*, 8(2), 2015, pp. 283–375.
44 Duncan Wright, ' Mid-Holocene Maritime Economy in the Western Torres Strait', *Archaeology in Oceania*, 46(1), 2011, pp. 23–7.
45 For example, see Lisa Yeomans, 'Evidence for Fishing with Remora Across the World and Archaeological Evidence from Southeast Arabia: A Case Study in Human-Animal Relations', *International Journal of Historical Archaeology*, 27(2), 2023, pp. 348–362.
46 Charles R Eastman, 'The Reversus: A Fishing Tale of Christopher Columbus', *Scientific Monthly*, 3(1), 1916, pp. 31–40.

47 John MacGillivray, *Narrative of the Voyage of HMS* Rattlesnake, *Volume 2* (T & W Boone, London, 1852), p. 21.
48 Walter E Roth, *Food: Its Search, Capture, and Preparation*, North Queensland Ethnography, Bulletin No. 3 (George A Vaughan, Acting Government Printer, Brisbane, 1901), p. 20.
49 Edmund J Banfield, *The Confessions of a Beachcomber* (T Fisher Unwin, London, 1908), p. 243.
50 Ibid., p. 244.
51 Haddon, *Reports of the Cambridge Anthropological Expedition to Torres Straits, Volume 4: Arts and Crafts* (Cambridge University Press, Cambridge, 1912), pp. 163–3.
52 Haddon, *Reports of the Cambridge Anthropological Expedition to Torres Straits, Volume 1*, pp. 65, 374.
53 Alfred C Haddon, 'Folktales', in Haddon, *Reports of the Cambridge Anthropological Expedition to Torres Straits, Volume 6: Sociology, Magic and Religion of the Eastern Islanders* (Cambridge University Press, Cambridge, 1908), pp. 1–63; Haddon, *Reports of the Cambridge Anthropological Expedition to Torres Straits, Volume 4*, p. 412.
54 Haddon, 'Magic and Religion', p. 364; Haddon, *Reports of the Cambridge Anthropological Expedition to Torres Straits, Volume 4*, p. 165.
55 Coralie D'Lima et al., 'Positive Interactions Between Irrawaddy Dolphins and Artisanal Fishers in the Chilika Lagoon of Eastern India Are Driven by Ecology, Socioeconomics, and Culture', *Journal of the Human Environment*, 43, 2014, pp. 614–24; Paulo C Simões-Lopes, ME Fabian & JO Menegheti, 'Dolphin Interactions with the Mullet Artisanal Fishing on Southern Brazil: A Qualitative and Quantitative Approach', *Revista Brasileira de Zoologia*, 15, 1998, pp. 709–26; Mauricio L Santos, Valéria M Lemos & João P Vieira, 'No Mullet, No Gain: Cooperation Between Dolphins and Cast Net Fishermen in Southern Brazil', *Zoologia*, 35, 2018.
56 David T Neil, 'Cooperative Fishing Interactions Between Aboriginal Australians and Dolphins in Eastern Australia', *Anthrozoös*, 15(1), 2002, pp. 3–18; Jay Hall, 'Fishing with Dolphins? Affirming a Traditional Aboriginal Fishing Story in Moreton Bay, SE Queensland', in Roger J

Coleman, Jeanette Covacevich & Peter Davie (eds), *Focus on Stradbroke* (Boolarong Publications, Brisbane, 1984), pp. 16–22.

57 John Curtis, *Shipwreck of the Stirling Castle* (George Virtue, London, 1838), p. 69.

58 Sarah Martin, *Eyre Peninsula and West Coast Aboriginal Fish Trap Survey* (South Australian Department of Environment and Planning, Adelaide, 1988), p. 36.

59 Robert H Mathews, 'Ethnological Notes on the Aboriginal Tribes of New South Wales and Victoria', *Journal and Proceedings of the Royal Society of New South Wales*, 38, 1904, pp. 252–3.

60 For a detailed discussion of the history of the killer whales of Eden, see Danielle Clode, *Killers in Eden: The True Story of the Killer Whales and Their Remarkable Partnership with the Whalers of Twofold Bay* (Allen & Unwin, Sydney, 2002).

61 Cited in Ian J McNiven & Damien Bell, 'Fishers and Farmers: Historicising the Gunditjmara Freshwater Fishery, Western Victoria', *La Trobe Journal*, 85, 2010, pp. 83–105.

62 Peter JF Coutts, Rudy K Frank & Phillip Hughes, *Aboriginal Engineers of the Western District, Victoria. Records of the Victorian Archaeological Survey 7* (VAS, Melbourne, 1978), p. 28.

63 Harry Lourandos, 'Aboriginal Settlement and Land Use in South Western Victoria: A Report on Current Field Work', *The Artefact*, 1(4), 1976, pp. 174–93.

64 Harry Lourandos, 'Swamp Managers of Southwestern Victoria', in Mulvaney & White (eds), *Australians to 1788*, pp. 293–307; Lourandos, 'Change or Stability?'

65 Heather Builth, 'Mt Eccles Lava Flow and the Gunditjmara Connection: A Landform for All Seasons', *Proceedings of the Royal Society of Victoria*, 116(1), 2004, pp. 163–82; Builth 2014.

66 Ian J McNiven et al., 'Dating Aboriginal Stone-Walled Fishtraps at Lake Condah, Southeast Australia', *Journal of Archaeological Science*, 39(2), 2012, pp. 268–86; Ian J McNiven et al., 'Phased Redevelopment of an Ancient Gunditjmara Fish Trap Over the Past 800 Years: Muldoons Trap Complex, Lake Condah, Southwestern Victoria', *Australian Archaeology*, 81, 2015, pp. 44–58.

67 Ashleigh J Rogers, 'Aquaculture in the Ancient World: Ecosystem Engineering, Integrated Landscapes, and the First Blue Revolution', *Journal of Archaeological Research*, in press.

68 Denis Rose, 'Community Aspirations for the World Heritage Listing of Budj Bim Lava Flow', *Historic Environment*, 26(2), 2014, pp. 98–100; Steve Brown et al., *Australia's Nomination of Budj Bim Cultural Landscape: World Heritage Nomination for Inscription in the UNESCO World Heritage List* (Department of Environment and Energy, Canberra, 2017); Anita Smith et al., 'Indigenous Knowledge and Resource Management as World Heritage Values: Budj Bim Cultural Landscape, Australia', *Archaeologies*, 15, 2019, pp. 285–313; Steve Brown, Anita Smith & Denis Rose, 'Aquaculture: Budj Bim Cultural Landscape, Australia', in Steve Brown & Cari Goetcheus (eds), *Routledge Handbook of Cultural Landscape Practice* (Routledge, London & New York, 2023), pp. 341–7.

69 Marilyn Truscott, 'Introduction', in *Australian Aboriginal Middens*, Australian Heritage Commission Bibliography Series. No. 10 (Australian Government Publishing Service, Canberra, 1994), p. iv; Sean Ulm, 'Coastal Foragers on Southern Shores: Marine Resource Use in Northeast Australia Since the Late Pleistocene', in Nuno F Bicho, Jonathan A Haws & Loren G Davis (eds), *Trekking the Shore: Changing Coastlines and the Antiquity of Coastal Settlement* (Springer Verlag, New York, 2011), pp. 441–61.

70 AD (Tam) Smith, 'Archaeological Expressions of Holocene Cultural and Environmental Change in Coastal Southeast Queensland', PhD thesis, School of Social Science, University of Queensland, 2016; Leslie Reeder-Myers et al., 'Indigenous Oyster Fisheries Persisted for Millennia and Should Inform Future Management', *Nature Communications*, 13(1), 2022, p. 2383.

71 Reeder-Myers et al., 'Indigenous Oyster Fisheries Persisted for Millennia and Should Inform Future Management'.

72 Annie Ross with members of the Quandamooka Aboriginal Land Council, 'Aboriginal Approaches to Cultural Heritage Management: A Quandamooka Case Study', in Sean Ulm, Ian Lilley & Annie Ross (eds), *Australian Archaeology '95: Proceedings of the 1995 Australian Archaeological*

Association Annual Conference, Tempus 6 (Anthropology Museum, University of Queensland, St Lucia, 1996), pp. 107–12.

73 Annie Ross, Shane Coghill & Brian Coghill, 'Discarding the Evidence: The Place of Natural Resources Stewardship in the Creation of the Peel Island Lazaret Midden, Moreton Bay, Southeast Queensland', *Quaternary International*, 385, 2015, pp. 177–90.

74 Lorraine M Woolley, *Histories of the Great Sandy Straits* (Pukpuk Publications, Portarlington, Vic., 2016).

6. OUTSIDE VISITOR OPPORTUNITIES

1 For example, see Henry Reynolds, *The Other Side of the Frontier: Aboriginal Resistance to the European Invasion of Australia* (UNSW Press, Randwick, 2006); Henry Reynolds, 'Settlers and Aborigines on the Pastoral Frontier', in BJ Dalton (ed.), *Lectures on North Queensland History* (History Department, James Cook University of North Queensland, Townsville, 1974), pp. 153–62.

2 Ann McGrath, *Born in the Cattle: Aborigines in Cattle Country* (Allen & Unwin, Sydney, 1987); Fred Cahir, *Black Gold: Aboriginal People on the Goldfields of Victoria*, 1850–1870 (ANU E Press & Aboriginal History Inc., Canberra, 2012); Alistair Paterson, 'The Archaeology of Agrarian Australia', in McNiven & David (eds), *The Oxford Handbook of the Archaeology of Indigenous Australia and New Guinea*; Julia Martínez & Adrian Vickers, *The Pearl Frontier: Indonesian Labor and Indigenous Encounters in Australia's Northern Trading Network* (University of Hawai'i Press, Honolulu, 2015).

3 Regina Ganter, *The Pearl-Shellers of Torres Strait: Resource Use, Development and Decline, 1860s–1960s* (Melbourne University Press, Carlton, Vic., 1994).

4 McGrath, *Born in the Cattle*; see also Paterson, 'The Archaeology of Agrarian Australia'.

5 Cahir, *Black Gold*; Martínez & Vickers, *The Pearl Frontier*.

6 Much has been written about Aboriginal women's exploitation at the hands of sealers – see, for example, Rebe Taylor, *Unearthed: The Aboriginal Tasmanians of Kangaroo Island* (Wakefield Press, Adelaide, 2002).

7 Lynette Russell, '"Tickpen", "Boro Boro": Aboriginal Economic Engagements in Early Melbourne', in Leigh Boucher & Lynette Russell (eds), *Settler Colonial Governance in Nineteenth Century Victoria* (ANU Press & Aboriginal History Inc., Canberra, 2015), pp. 27–46; Lynette Russell & Leigh Boucher, '"Soliciting Sixpences from Township to Township": Moral Dilemmas in Mid-Nineteenth-Century Melbourne', *Postcolonial Studies*, 15(2), 2012, pp. 149–65.

8 *The Age*, 10 December 1883, p. 5.

9 Richard Broome, *Aboriginal Victorians: A History Since 1800* (Allen & Unwin, Sydney, 2005).

10 AGL Shaw, *Victoria Before Separation: A History of the Port Phillip District* (Miegunyah Press, Carlton South, Vic., 1996), pp. 115–16.

11 For a comprehensive history and analysis of this 'Treaty', see Bain Attwood with Helen Doyle, *Possession: Batman's Treaty and the Matter of History* (Miegunyah Press, Melbourne, 2009). See also James Boyce, *1835: The Founding of Melbourne & the Conquest of Australia* (Black Inc., Melbourne, 2011); James Bonwick, *John Batman, the Founder of Victoria* (Wren, Melbourne, 1868).

12 Pauline Jones (ed.), *Historical Records of Victoria, Volume 1: Beginnings of Permanent Government* (Victorian Government Printer, Melbourne, 1982), p. 102.

13 Edward M Curr, *Recollections of Squatting in Victoria, Then Called Port Phillip District (from 1841 to 1851)* (George Robertson, Melbourne, 1883), p. 21.

14 Broome, *Aboriginal Victorians*, pp. 148–9.

15 Cahir, *Black Gold*, pp. 47–56.

16 Quoted in Marie Hansen Fels, *'I Succeeded Once': The Aboriginal Protectorate on the Mornington Peninsula, 1839–1840*, Aboriginal History Monograph 22 (ANU ePress, Canberra, 2011), p. 112.

17 Quoted in Penelope Edmonds, *Urbanizing Frontiers: Indigenous Peoples and Settlers in Pacific Rim Cities* (UBC Press, Vancouver, 2010), p. 46. The extensive diary entry includes a delicate pen-and-ink sketch of an Aboriginal man and woman and a small dog ('William Adeney Diary Sketch, Aborigines in Melbourne, 1843', SLV, MS 8520, pp. 296–307).

18 Alan Pope, 'Aboriginal Adaptation to Early Colonial Labour Markets: The South Australian Experience', *Labour History*, 54, 1988, pp. 1–15. See also Broome, *Aboriginal Victorians*, pp. 148–9; R Castle & J Hagan, 'Centuries of Aboriginal Unemployment in NSW', *Modern Unionist* 8, 1983. For an excellent discussion of the colonial representations of Indigenous people as 'naturally idle', see Syed Hussein Alatas, *The Myth of the Lazy Native* (F Cass, London, 1977), p. 9.
19 Edmonds, *Urbanizing Frontiers*, p. 88.
20 Today the site where the Merri Creek Aboriginal School sat is dominated by the footings of the Eastern Freeway bridge, constructed in the 1970s. No remains have been located of the school (or the nearby Aboriginal Protectorate Station): it is most likely that the freeway construction and the redirection of the Yarra River destroyed all evidence.
21 Lucy A Edgar, *Among the Black Boys: Being the History of an Attempt at Civilising Some Young Aborigines of Australia* (Emily Faithful, London, 1865), pp. 48–53.
22 Edgar, *Among the Black Boys*, p. 49.
23 Ian Clark (ed.), *The Journals of George Augustus Robinson, Chief Protector, Port Phillip Aboriginal Protectorate, Volume 5* (Heritage Matters, Melbourne, 2000), entry for 5 June 1849.
24 Edgar, *Among the Black Boys*, p. 54.
25 J Cruickshank & M McMillan, 'Lawful Conduct, Aboriginal Protection and Land in Victoria, 1859–1869', in Samuel Furphy & Amanda Nettelbeck (eds), *Aboriginal Protection and Its Intermediaries in Britain's Antipodean Colonies* (Routledge, New York, 2020), pp. 194–211.
26 Edmonds, *Urbanizing Frontiers*, p. 88.
27 For a detailed examination of the whaling and sealing industry and Indigenous participation, see Lynette Russell, *Roving Mariners: Australian Aboriginal Whalers and Sealers in the Southern Oceans, 1790–1870* (State University of New York Press, New York, 2012).
28 Michael Nash, *The Bay Whalers: Tasmania's Shore-Based Whaling Industry* (Navarine Publishing, Canberra, 2003), p. 91.
29 Lyndall Ryan, *The Aboriginal Tasmanians*, 2nd edn (Allen & Unwin, St Leonards, NSW, 1996), p. 214.

30 For example, Erin Parke, 'Discovery of Illegal Indonesian Fishing Camp on Kimberley Coast Prompts Biosecurity, Border Concerns', *ABC News*, 29 August 2022.

31 Makassan tradition holds that only men travelled to Australia; however, some of these men 'married' Aboriginal women, and some of those women travelled to Makassar with their husbands.

32 Campbell C Macknight, *The Voyage to Marege': Macassan Trepangers in Northern Australia* (Melbourne University Press, Carlton, Vic., 1976); Peter Grave & Ian J McNiven, 'Geochemical Identification of Asian Stoneware Jars from Torres Strait, Northeast Australia', *Journal of Archaeological Science*, 40, 2013, pp. 4538–51.

33 Nick Evans, 'Macassan Loanwords in Top End Languages', *Australian Journal of Linguistics*, 12(1), 1992, pp. 45–91; Campbell C Macknight, 'Harvesting the Memory: Open Beaches in Makassar and Arnhem Land', in Peter Veth, Peter Sutton & Margo Neale (eds), *Strangers on the Shore: Early Coastal Contacts in Australia* (National Museum of Australia, Canberra, 2008), pp. 133–47.

34 Kenneth Morgan (ed.), *Australia Circumnavigated: The Voyage of Matthew Flinders in HMS Investigator, 1801–1803, Volume II* (Hakluyt Society, London, 2015).

35 McNiven, 'Beyond Bridge and Barrier'.

36 Scott RA Mitchell, 'Culture Contact and Indigenous Economies on the Cobourg Peninsula, Northwestern Arnhem Land', PhD dissertation, Northern Territory University, 1994, p. 113.

37 Ronald M Berndt & Catherine H Berndt, *Arnhem Land: Its History and Its People* (CH Cheshire, Melbourne, 1954); Annie Clarke, 'The "Moormans trousers": Macassan and Aboriginal Interactions and the Changing Fabric of Indigenous Social Life', in Sue O'Connor & Peter Veth (eds), *East of Wallace's Line: Studies of Past and Present Maritime Cultures of the Indo-Pacific* (AA Balkema, Rotterdam, 2000), pp. 315–35; Macknight, *The Voyage to Marege'*.

38 Chris Urwin, Lynette Russell & Lily Yulianti, 'Cross-Cultural Interaction Across the Arafura and Timor Seas: Aboriginal People and Macassans in Northern Australia', in McNiven & David, *The Oxford Handbook of the*

Archaeology of Indigenous Australia and New Guinea. Urwin et al., 2016, using a 25-year generational spacing, estimate up to nine generations.

39 Clarke, 'The "Moormans trousers"'.

40 Daryl Wesley & Mirani Litster, '"Small, Individually Nondescript and Easily Overlooked": Contact Beads from Northwest Arnhem Land in an Indigenous-Macassan-European Hybrid Economy', *Australian Archaeology*, 80, 2015, pp. 1–16.

41 George W Earl, *Enterprise in Tropical Australia* (Madden & Malcolm, London, 1846), p. 77; Mitchell, 1994, Table 6-1, for a summary.

42 Evans, 'Macassan Loanwords in Top End Languages'; Campbell C Macknight, 'Macassans and Aborigines', *Oceania*, 42(4), 1972, pp. 283–321.

43 George Chaloupka, 'Praus in Marege: Makassan Subjects in Aboriginal Rock Art of Arnhem Land, Northern Territory, Australia', *Anthropologie*, 14(1–2), 1996, pp. 131–42.

44 Sally K May et al., 'Painted Ships on a Painted Arnhem Land Landscape', *The Great Circle: Journal of the Australian Association for Maritime History*, 35(2), 2013, pp. 83–102; Sally K May et al., 'The World from Malarrak: Depictions of Southeast Asian and European Subjects in Rock Art from the Wellington Range, Australia', *Australian Aboriginal Studies*, 1, 2013, pp. 45–56; Paul SC Taçon & Sally K May, 'Ship Shape: An Exploration of Maritime-Related Depictions in Indigenous Rock Art and Material Culture', *The Great Circle*, 35(2), 2013, pp. 7–15.

45 Sally K May et al., 'The Missing Macassans: Indigenous Sovereignty, Rock Art and the Archaeology of Absence', *Australian Archaeology*, 87(2), 2021, pp. 127–43.

46 Gurrumul's family have given specific permission for his name to written and spoken; in Yolŋu culture it is generally the custom to avoid using the names of deceased people. According to Skinnyfish Music, Gurrumul's family were concerned that his legacy and music should not be forgotten, so they made the extraordinary decision to allow his name, image and voice to be used.

7. REPURPOSING TECHNOLOGIES, ART AND COSMOLOGIES

1 Ian J McNiven & Myles Russell-Cook, 'Seeing What They Saw: Harden Sidney Melville's *Torres Strait canoe and five men at the site of a wreck on the Sir Charles Hardy Islands, off Cape Grenville, North East Australia, 1874*', *NGV Magazine*, 25, 2020, pp. 14–17.

2 Grave & McNiven, 'Geochemical Identification of Asian Stoneware Jars from Torres Strait, Northeast Australia'.

3 Nicolas Peterson, 'Ethno-Archaeology in the Australian Iron Age', in Gale de G Sieveking, Ian H Longworth & KE Wilson (eds), *Problems in Economic and Social Archaeology* (Duckworth, London, 1976), pp. 265–75.

4 Harry Allen, 'Thomson's Spears: Innovation and Change in Eastern Arnhem Land Projectile Technology', in Yasmine Musharbash & Marcus Barber (eds), *Ethnography and the Production of Anthropological Knowledge: Essays in Honour of Nicolas Peterson* (ANU ePress, Canberra, 2011), pp. 69–88.

5 Macknight, *The Voyage to Marege'*.

6 Macknight, *The Voyage to Marege'*.

7 Joseph B Jukes, *Narrative of the Surveying Voyage of HMS* Fly, *Volume 1* (T & W Boone, London, 1847), p. 359.

8 Scott Mitchell, 'Foreign Contact and Indigenous Exchange Networks on the Cobourg Peninsula, North Western Arnhem Land', *Australian Aboriginal Studies*, 2, 1995, pp. 44–8.

9 Haddon, *Reports of the Cambridge Anthropological Expedition to Torres Straits, Volume 1*, pp. 93, 114, 325; Moore (ed.), *Islanders and Aborigines at Cape York*, pp. 184, 253.

10 Reynolds, *The Other Side of the Frontier*; Peter Veth & Susan O'Connor, 'Archaeology, Claimant Connection to Sites, and Native Title: Employment of Successful Categories of Data with Specific Comments on Glass Artifacts', *Australian Aboriginal Studies*, 1, 2005, pp. 2–15; Simon Munt & Timothy D Owen, 'Aboriginal Uses for Introduced Glass, Ceramic and Flint from the Former Schofields Aerodrome, Western Sydney (Darug Country), New South Wales', *Australian Archaeology*, 88(1), 2022, pp. 49–64; Yinika Perston et al., 'Flaked Glass Artifacts from Nineteenth–Century Native Mounted Police Camps in

Queensland, Australia', *International Journal of Historical Archaeology*, 26, 2022, pp. 789–822.

11 Nathan Wolski, 'Brushing Against the Grain: Excavating for Aboriginal-European Interaction on the Colonial Frontier in Western Victoria, Australia', PhD thesis, University of Melbourne, 2000. See also Tom H Loy & Nathan Wolski, 'On the Invisibility of Contact: Residue Analyses on Aboriginal Glass Artefacts from Western Victoria', *The Artefact*, 22, 1999, pp. 65–73.

12 Veth & O'Connor, 'Archaeology, Claimant Connection to Sites, and Native Title'.

13 Jones, *Ochre and Rust*, p. 126.

14 Edward C Stirling, *Report on the Work of the Horn Scientific Expedition to Central Australia, Part IV: Anthropology*, Baldwin Spencer (ed.) (Dulau & Co. , London, 1896), p. 95.

15 Kim Akerman, Richard Fullagar & Annelou van Gijn, 'Weapons and Wunan: Production, Function and Exchange of Kimberley Points', *Australian Aboriginal Studies*, 1, 2002, pp. 13–42; Kim Akerman, 'Material Culture and Trade in the Kimberleys Today', in Ronald M Berndt (ed.), *Aborigines of the West* (University of Western Australia Press, Nedlands, 1979), pp. 243–51.

16 Akerman, Fullagar & van Gijn, 'Weapons and Wunan'.

17 Rodney Harrison, 'An Artefact of Colonial Desire? Kimberley Points and the Technologies of Enchantment', *Current Anthropology*, 47(1), 2006, pp. 63–88.

18 Rodney Harrison, 'Archaeology and the Colonial Encounter: Kimberley Spearpoints, Cultural Identity and Masculinity in the North of Australia', *Journal of Social Archaeology*, 2(3), 2002, pp. 352–77; Rodney Harrison, 'Kimberley Points and Colonial Preference: New Insights into the Chronology of Pressure Flaked Point Forms from the Southeast Kimberley, Western Australia', *Archaeology in Oceania*, 39(1), 2004, pp. 1–11; Harrison, 'An Artefact of Colonial Desire?'

19 Kim Akerman with John Stanton, *Riji and Jakuli: Kimberley Pearl Shell in Aboriginal Australia*, Monograph Series 4 (Northern Territory Museum of Arts and Sciences, Darwin, 1994). See also McCarthy,

'"Trade" in Aboriginal Australia, and "Trade" Relationships with Torres Strait, New Guinea and Malaya', *Oceania*, 9(4), 1939, pp. 405–39; 10(1), 1940, pp. 80–104; 10(2), 1940, pp. 171–95; Daniel S Davidson, 'The Interlocking Key Design in Aboriginal Australian Decorative Art', *Mankind*, 4(3), 1949, pp. 85–98; Tanya Edwards & Sarah Yu (eds), *Lustre: Pearling & Australia* (Western Australian Museum, Perth, 2018).

20 Akerman with Stanton, *Riji and Jakuli*, p. 15.

21 Akerman with Stanton, *Riji and Jakuli*, p. 15.

22 George Grey, *Journals of Two Expeditions of Discovery, Volume 1* (T & W Boone, London, 1841), p. 203.

23 Peter A Elkin, 'Grey's Northern Kimberley Cave-Paintings Re-Found', *Oceania*, 19(1), 1948, pp. 1–15.

24 Elkin, 'Grey's Northern Kimberley Cave-Paintings Re-Found', p. 6.

25 Catherine Frieman & Sally K May, 'Navigating Contact: Tradition and Innovation in Australian Contact Rock Art', *International Journal of Historical Archaeology*, 24, 2020, pp. 342–66.

26 Frieman & May, 'Navigating Contact'.

27 Sue Ryan, 'Ghost Nets Moa Island Puppets', *Textile Fibre Forum*, 30(2), 2011, pp. 46–9; Sue Ryan, 'The Ghost Net Art Project', *Artlink*, 32(2), 2012, pp. 108–12; G Le Roux, 'Transforming Representations of Marine Pollution: for a New Understanding of the Artistic Qualities and Social Values of Ghost Nets', *AnthroVision: Vaneasa Online Journal*, 4(1), 2016; Riki Gunn, Britta Denise Hardesty & James Butler, 'Tackling "Ghost Nets": Local Solutions to a Global Issue in Northern Australia', *Ecological Management & Restoration*, 11(2), 2010, pp. 88–98.

28 'Erub Arts: An Art Centre with a Difference', Erub Arts website, 2018.

29 Myles Russell-Cook, *Maree Clarke: Ancestral Memories* (NGV, Melbourne, 2021).

30 Myles Russell-Cook, 'Artists Maree Clark and Lyn-Al Young's Unique Collaboration Celebrates Their Indigenous Heritage', *Vogue Australia*, 16 December 2019.

31 Treena Clark & Peter McNeil, '"Cultural Expression Through Dress": Towards a Definition of First Nations Fashion', *The Conversation*, 22 March 2023.

32 Treena Clark et al., '"I Want to Create Change; I Want to Create Impact": Personal-Activism Narratives of Indigenous Australian Women Working in Public Relations', *Public Relations Review*, 48(1), 2022.
33 Peter Willis, 'Riders in the Chariot: Aboriginal Conversion to Christianity at Kununurra', in Tony Swain & Deborah Bird Rose, *Aboriginal Australians and Christian Missions* (Australian Association for the Study of Religions, Bedford Park, SA, 1988), pp. 308–20.
34 Peta Stephenson, 'Indigenous Australia's Pilgrimage to Islam', *Journal of Intercultural Studies*, 32, 2011, pp. 261–77.
35 Garry Deverell, *Gondwana Theology: A Trawloolway Man Reflects on Christian Faith* (Morning Star Publishing, Reservoir, Vic., 2018).
36 Deverell, *Gondwana Theology*, pp. 19, 40–1.
37 Bain Attwood, *William Cooper: An Aboriginal Life Story* (Melbourne University Press, Carlton, Vic., 2021).
38 Swain & Rose (eds), *Aboriginal Australians and Christian Missions*; Carolyn Schwarz & Françoise Dussart, 'Christianity in Aboriginal Australia Revisited', *Australian Journal of Anthropology*, 21(1), 2010, pp. 1–13; Lynne Hume, 'The Dreaming in Contemporary Aboriginal Australia', in Graham Harvey (ed.), *Indigenous Religions: A Companion* (Cassell, London, 2000), pp. 125–38.
39 Abu Bakr Sirajuddin Cook & Salih Yucel, 'Australia's Indigenous Peoples and Islam: Philosophical and Spiritual Convergences Between Belief Structures', *Comparative Islamic Studies*, 12(1–2), 2016, pp. 165–85.
40 Peter Toner, 'Ideology, Influence and Innovation: The Impact of Macassan Contact on Yolngu music', *Perfect Beat: The Pacific Journal of Research into Contemporary Music and Popular Culture*, 5(1), 2000, pp. 22–41.
41 Ian S McIntosh, *Between Two Worlds: Essays in Honour of the Visionary Aboriginal Elder, David Burrumarra* (Dog Ear Publishing, Indianapolis, 2015).
42 Jane Balme, Sue O'Connor & Stewart Fallon, 'New Dates on Dingo Bones from Madura Cave Provide Oldest Firm Evidence for Arrival of the Species in Australia', *Scientific Reports*, 8(1), 2018.
43 Bradley Smith & Carla Litchfield, 'A Review of the Relationship Between Indigenous Australians, Dingoes (*Canis dingo*) and Domestic Dogs (*Canis familiaris*)', *Anthrozoös*, 22(2), 2009, pp. 111–28.

44 Annette Hamilton, 'Aboriginal Man's Best Friend?', *Mankind*, 8, 1972, pp. 287–95.
45 Deborah Bird Rose, *Dingo Makes Us Human: Life and Land in an Aboriginal Australian Culture* (Cambridge University Press, Cambridge, 1992), p. 176.
46 Erich Kolig, 'Aboriginal Man's Best Foe', *Mankind*, 9, 1973, pp. 122–4; Mervyn Meggitt, 'Australian Aborigines and Dingoes', in Anthony Leeds & Andrew P Vayda (eds), *Man, Culture and Animals* (American Association for the Advancement of Science, Washington, DC, 1965), pp. 7–26.
47 Betty Meehan, Rhys Jones & Annie Vincent, 'Gulu-Kula: Dogs in Anbarra Society, Arnhem Land', *Aboriginal History*, 23, 1999, pp. 83–106.
48 Smith & Litchfield, 'A Review of the Relationship Between Indigenous Australians, Dingoes (*Canis dingo*) and Domestic Dogs (*Canis familiaris*)'.
49 Rose, *Dingo Makes Us Human*.
50 Ian McIntosh, 'Why the Dingo Ate Its Master', *Australian Folklore*, 14, 1999, pp. 183–9.
51 Robert Gunn, Ray L Whear & Leigh C Douglas, 'A Dingo Burial from the Arnhem Land Plateau', *Australian Archaeology*, 71(1), 2010, pp. 11–16.
52 Colin Pardoe, 'Dogs Changed the World', *National Dingo News*, Spring 1996, pp. 19–29.
53 Douglas W Bird et al., 'Megafauna in a Continent of Small Game: Archaeological Implications of Martu Camel Hunting in Australia's Western Desert', *Quaternary International*, 297, 2013, pp. 155–66.
54 David Trigger, 'Indigeneity, Ferality, and What "Belongs" in the Australian Bush: Aboriginal Responses to "Introduced" Animals and Plants in a Settler-Descendant Society', *Journal of the Royal Anthropological Institute*, 14(3), 2008, pp. 628–46.
55 Mitchell Rolls, 'Black Is Not Green', *Australian Studies*, 18(1), 2003, pp. 41–65.
56 Andrea Gaynor, 'Report on the History of the Arrival of the Feral Cat Population in Western Australia', *CALM Science*, 3, 2000, pp. 149–79; Gillian Hutcherson, *Gong-wapitja: Women and Art from Yirrkala* (Aboriginal Studies Press, Canberra, 1998).

57 Yanyuwa Families and John Bradley, *Wuka nya-nganunga li-Yanyuwa li-Anthawirriyarra = Language for Us, the Yanyuwa Saltwater People: A Yanyuwa Encyclopedia, Volume 2* (Australian Scholarly Publishing, North Melbourne, 2017), pp. 101, 240.
58 C Pavey, *National Recovery Plan for the Greater Bilby* (Northern Territory Department of Natural Resources, Environment and the Arts, Alice Springs, 2006).
59 'Managing Feral Animals', Central Land Council website, 2023.

8. CHANGING AUSTRALIA

1 Patrick D Nunn, *The Edge of Memory: Ancient Stories, Oral Tradition and the Post-Glacial World* (Bloomsbury Sigma, London & New York, 2018), pp. 201–2.
2 *Wunungu Awara* is a Yanyuwa term for a strong and healthy, vital place.
3 Ian J McNiven, 'Canoes of Mabuyag and Torres Strait', *Memoirs of the Queensland Museum: Culture*, 8(1), 2015, pp. 127–207; Ian J McNiven, 'Torres Strait Canoes as Social and Predatory Object-Beings', in Eleanor Harrison-Buck & Julie A Hendon (eds), *Relational Identities and Other-Than-Human Agency in Archaeology* (University of Colorado Press, Denver, 2018), pp. 167–96; McNiven & Russell-Cook, 'Seeing What They Saw'.
4 Ian J McNiven, Thomas Chandler & Michael Neylan, 'Recreating a Torres Strait Canoe', in Mitchell & Ridgway (eds), *Connections Across the Coral Sea*, pp. 22–7.
5 Liam M Brady, *Pictures, Patterns and Objects: Rock-Art of the Torres Strait Islands, Northeastern Australia* (Australian Scholarly Publishing, North Melbourne, 2010).
6 McNiven, 'Canoes of Mabuyag and Torres Strait'.
7 Quoted in Niel Gunson, 'An Interview with David Unaipon', *Aboriginal History*, 23, 1999, pp. 111–16.
8 'An Aboriginal Genius', *Daily Herald*, 1 June 1914, p. 9.
9 'An Aboriginal Genius'.
10 'Patent specification. Mechanical motion. Sheep shears. No. 15,624, 1909. D. Unaipon, SA', in *Australian Official Journal of Patents*, 1910, p. 910.
11 'David Ngunaitponi (Unaipon): The Brilliant Inventor', AIATSIS website, n.d.

12 David Unaipon, *Legendary Tales of the Australian Aborigines*, eds Stephen Muecke & Adam Shoemaker (Miegunyah Press, Carlton, Vic., 2001 [1924–25]); 'David Ngunaitponi (Unaipon): The Brilliant Orator and Writer', AIATSIS website, n.d.

13 'Family Demands Payout for $50 Note Image of David Unaipon', *Daily Telegraph*, 28 November 2008. Recent conversation with Ngarrindjeri community members indicate that the concern over Unaipon's appropriation on the fifty-dollar note continues to this day.

14 Kathy Swan, 'David Unaipon Inspires Theatre Production', *ABC News*, 14 June 2015. Quote in Mathew Rimmer, 'Preface: The legacy of David Unaipon', in *Indigenous Intellectual Property: A Handbook of Contemporary Research* (Edward Elgar Publishing, Cheltenham, Vic., 2015), pp. xxi–xxviii.

15 In both the play and in the documentary *Bastardy* (directed by Amiel Courtin-Wilson, 2008), Uncle Jack refers to his craving for heroin as 'chasing the dragon'.

16 'About the Yoorrook Justice Commission', Yoorrook Justice Commission website, 2023.

17 Thomas Crofts & Tanya Mitchell, 'Prohibited Behaviour Orders and Indigenous Overrepresentation in the Criminal Justice System', *Current Issues in Criminal Justice*, 23(2), 2011, pp. 277–85.

18 National Inquiry into the Separation of Aboriginal and Torres Strait Islander Children from their Families, *Bringing Them Home: Report of the National Inquiry into the Separation of Aboriginal and Torres Strait Islander Children from their Families* (Australian Human Rights Commission, Sydney, 1997).

19 See Kate Auty & Lynette Russell, *Hunt Them, Hang Them: The Tasmanians in Port Phillip 1842* (Justice Press, Melbourne, 2016).

20 Kate Auty & Daniel Briggs, 'Koori Court Victoria: Magistrates Court (Koori Court) Act 2002', *Law Text Culture*, 8(8), 2004, pp. 7–38.

21 Mark Harris, 'The Koori Court and the Promise of Therapeutic Jurisprudence', *eLaw Journal: Murdoch University Electronic Journal of Law* 1, 2007.

22 Auty & Briggs, 'Koori Court Victoria', p. 23.

23 Inspired by the success of the Victorian system, New South Wales developed its own Koori courts – see Evarn J Ooi & Sara Rahman, *The Impact of the NSW Youth Koori Court on Sentencing and Re-offending Outcomes*, Crime and Justice Bulletin 248 (NSW Bureau of Crime Statistics and Research, Parramatta, 2022).

24 For an excellent discussion on fire and burning, see Gammage & Pascoe, *Country: Future Fire, Future Farming*. See also Penny Olsen & Lynette Russell, *Australia's First Naturalists: Indigenous Peoples' Contribution to Early Zoology* (National Library of Australia, Canberra, 2019).

25 J Russell-Smith et al., 'Managing Fire Regimes in North Australian Savannas: Applying Aboriginal Approaches to Contemporary Global Problems', *Frontiers in Ecology and Environment*, 11(s1), 2013, pp. 55–63; LA Pastro et al., 'Burning for Biodiversity or Burning Biodiversity? Prescribed Burn vs. Wildfire Impacts on Plants, Lizards, and Mammals', *Ecological Applications*, 21, 2011, pp. 3238–53; E Young & H Ross, 'Using the Aboriginal Rangelands: "Insider" Realities and "Outsider" Perceptions', *Rangeland Journal*, 16, 1994, pp. 184–97; Olsen & Russell, *Australia's First Naturalists*, pp. 179–81.

26 Greg Lehman, 'Turning Back the Clock: Fire, Biodiversity, and Indigenous Community Development in Tasmania', in R Baker et al. (eds), *Working on Country: Contemporary Indigenous Management of Australia's Lands and Coastal Regions* (Oxford University Press, South Melbourne, 2001), pp. 308–19.

27 D Smyth et al., *Indigenous Land and Sea Management: A Case Study*, report prepared for the Australian Government Department of Sustainability, Environment, Water, Population and Communities on behalf of the State of the Environment 2011 Committee (DSEWPaC, Canberra, 2011).

28 S Woenne-Green et al., *Competing Interests: Aboriginal Participation in National Parks and Conservation Reserves in Australia: A Review* (Melbourne: Australian Conservation Foundation, 1994); M Bomford & J Caughley (eds), *Sustainable Use of Wildlife by Aboriginal Peoples and Torres Strait Islanders* (Australian Government Publishing Service, Canberra, 1996); T Corbett, M Lane & C Clifford, *Achieving Indigenous Involvement in Management of Protected Areas: Lessons from*

Recent Australian Experiences, Aboriginal Politics and Public Sector Management Research Paper no. 5 (Centre for Australian Public Sector Management, Griffith University, Nathan, Qld, 1998).

29 B Gilligan, *The Indigenous Protected Areas Programme: 2006 Evaluation* (Department of the Environment and Heritage, Canberra, 2006).

30 SVA Consulting, *Consolidated Report on Indigenous Protected Areas Following Social Return on Investment Analyses* (National Indigenous Australians Agency, Department of the Prime Minister & Cabinet, Canberra, 2016); Synergies Economic Consulting, *Working for Our Country: A Review of the Economic and Social Benefits of Land and Sea Management* (Pew Charitable Trusts, Philadelphia, PA, 2015).

31 JCZ Woinarski et al., *The Action Plan for Australian Mammals 2012* (CSIRO Publishing, Collingwood, Vic., 2014); Department of Climate Change, Energy, the Environment and Water, 'Species Profile and Threats Database', DCCRRW website; H Marsh, A Grech & R Hagihara, *Aerial Survey of Torres Strait to Evaluate the Efficacy of an Enforced and Possibly Extended Dugong Sanctuary as One of the Tools for Managing the Dugong Fishery* (Australian Marine Mammal Centre and the Torres Strait Regional Authority, Canberra, 2011); J Butler et al., 'Integrating Traditional Ecological Knowledge and Fisheries Management in the Torres Strait, Australia: The Catalytic Role of Turtles and Dugong as Cultural Keystone Species', *Ecology and Society*, 17(4), 2012; John C Altman & Sue Jackson, 'Indigenous Land and Sea Management', in D Lindenmayer et al. (eds), *Ten Commitments Revisited: Securing Australia's Future Environment* (CSIRO Publishing, Collingwood, Vic., 2014), pp. 207–16.

32 'Yirralka Rangers: Who We Are', Laynhapuy Homelands Aboriginal Corporation website, 2017; 'Species Profile and Threats Database'.

33 'Li-Anthawirriyarra Sea Ranger Unit', Mabunji Aboriginal Resource Centre website, 2016. See also John J Bradley, '"Li-Maramanja": Yanyuwa Hunters of Marine Animals in the Sir Edward Pellew Group, Northern Territory', *Records of the South Australian Museum*, 25(1), 1991, pp. 91–110.

34 McNiven & Bedingfield, 'Past and Present Marine Mammal Hunting Rates and Abundances'.

35 Chris Urwin et al., 'Hearing the Evidence: Using Archaeological Data to Analyse the Long-Term Impacts of Dugong (*Dugong dugon*) Hunting on Mabuyag, Torres Strait, over the Past 1000 Years', *Australian Archaeology*, 82(3), 2016, pp. 201–17.

36 Helene Marsh, 'Dugongs Are Safer in Torres Strait than Townsville', *The Conversation*, 10 May 2013.

9. LOOKING FORWARD

1 For a detailed exploration of this shameful history, see McNiven & Russell, *Appropriated Pasts*.

2 See various chapters in Claire Smith & H Martin Wobst (eds), *Decolonizing Archaeological Theory and Practice* (Routledge, New York, 2005); Jane Lydon & Uzma Z Rizvi (eds), *Handbook of Postcolonial Archaeology* (Left Coast Press, Walnut Creek, CA, 2010). See also McNiven & Russell, *Appropriated Pasts*, Chapter 8; Chris Wilson, 'Indigenous Research and Archaeology: Transformative Practices in/with/for the Ngarrindjeri Community', *Archaeologies*, 3(3), 2007, pp. 220–334; Ian J McNiven, 'Theoretical Challenges of Indigenous Archaeology: Setting an Agenda', *American Antiquity*, 81(1), 2016, pp. 27–41; Kellie Pollard et al., 'Indigenous Views on the Future of Public Archaeology in Australia: A Conversation that Did Not Happen', *AP: Online Journal in Public Archaeology*, 10, 2020, pp. 31–52.

3 The phrase 'Nothing about us without us' was first used by disability activists. It has subsequently been widely taken up by Indigenous people around the world. See James Charlton, *Nothing About Us Without Us: Disability Oppression and Empowerment* (University of California Press, Berkeley, 2000).

4 Asmi Wood, 'Treaty Making (Makarrata) and an "Invisible" People: Seeking a Just Peace After "Conflict"', in K Te Maiharoa, M Ligaliga & H Devere (eds), *Decolonising Peace and Conflict Studies through Indigenous Research* (Springer, Singapore, 2022), pp. 231–68.

5 William EH Stanner, 'The Great Australian Silence', in *The 1968 Boyer Lectures: After the Dreaming* (ABC Enterprises, Sydney, 1969), pp. 18–29.

FURTHER READING

Arthur, William & Frances Morphy, *Macquarie Atlas of Indigenous Australia*, 2nd edn (Macmillan Australia, Sydney, 2019).

Bashford, Alison & Stuart Macintyre, *The Cambridge History of Australia* (Cambridge University Press, Melbourne, 2013).

Bradley, John with Yanyuwa Families, *Singing Saltwater Country: Journey to the Songlines of Carpentaria* (Allen & Unwin, Sydney, 2010).

Broome, Richard, *Aboriginal Australians: A History Since 1788* (Allen & Unwin, Sydney, 2019).

Cane, Scott, *First Footprints: The Epic Story of the First Australians* (Allen & Unwin, Sydney, 2013).

Coleman, Claire G, *Lies, Damned Lies: A Personal Exploration of the Impact of Colonisation* (Ultimo Press, Ultimo, NSW, 2021).

Davis, Megan & George Williams, *Everything You Need to Know About the Uluru Statement from the Heart* (UNSW Press, Sydney, 2021).

Foley, Fiona, *Biting the Clouds: A Badtjala Perspective on the Aboriginals Protection and Restriction of the Sale of Opium Act, 1897* (University of Queensland Press, St Lucia, 2020).

Grant, Stan, *Talking to My Country* (HarperCollins, Sydney, 2016).

Griffiths, Billy, *Deep Time Dreaming: Uncovering Ancient Australia* (Black Inc., Melbourne, 2018).

Heiss, Anita, *Growing Up Aboriginal in Australia* (Black Inc., Melbourne, 2018).

Irish, Paul, *Hidden in Plain View: The Aboriginal People of Coastal Sydney* (NewSouth Publishing, Sydney, 2017).

Janke, Terri, *True Tracks: Respecting Indigenous Knowledge And Culture* (NewSouth Publishing, Sydney, 2021).

Jones, Philip, *Ochre and Rust: Artefacts and Encounters on Australian Frontiers* (Wakefield Press, Kent Town, SA, 2007).

Langton, Marcia, *Welcome to Country: An Introduction to Our First People for Young Australians* (Hardie Grant, Richmond, Vic., 2019).

Nakata, Martin, *Disciplining the Savages: Savaging the Disciplines* (Aboriginal Studies Press, Canberra, 2007).

Nunn, Patrick, *The Edge of Memory: Ancient Stories, Oral Tradition and the Post-Glacial World* (Bloomsbury Sigma, London, 2018).
Rose, Deborah Bird, *Dingo Makes Us Human: Life and Land in an Aboriginal Australian Culture* (Cambridge University Press, Melbourne, 1992).
Watego, Chelsea, *Another Day in the Colony* (University of Queensland Press, St Lucia, 2021).

INDEX

Note: Page numbers in **bold** refer to captions or images.

The best of both worlds

TITLES IN THE FIRST KNOWLEDGES SERIES

SONGLINES
Margo Neale & Lynne Kelly
(2020)

DESIGN
Alison Page & Paul Memmott
(2021)

COUNTRY
Bill Gammage & Bruce Pascoe
(2021)

ASTRONOMY
Karlie Noon & Krystal De Napoli
(2022)

PLANTS
Zena Cumpston, Michael-Shawn Fletcher & Lesley Head
(2022)

LAW
Marcia Langton & Aaron Corn
(2023)

INNOVATION
Ian J McNiven & Lynette Russell
(2023)

MEDICINE
Shawana Andrews, Sandra Eades & Fiona Stanley
(2024)

SEASONS
Patricia Marrfurra McTaggart,
Lorraine Williams & Emma Woodward
(2024)

CEREMONY
Wesley Enoch & Georgia Curran
(2025)

POLITICS
Morgan Brigg & Mary Graham
(2025)

Published in conjunction with the National Museum of Australia
and supported by the Australia Council for the Arts.

THE FIRST SIX FIRST KNOWLEDGES BOOKS ARE NOW AVAILABLE IN A BOX SET

The First Knowledges series offers an introduction to Indigenous knowledges in vital areas and their application to the present day and the future.

Out in November 2023

Visit us at: thamesandhudson.com.au
Follow us on Instagram: @thamesandhudsonau
And Facebook: @thameshudsonaustralia